SUPER SIMPLE

ANATOMY & PHYSIOLOGY

THE ULTIMATE LEARNING TOOL

SUPER SIMPLE

ANATOMY & PHYSIOLOGY

THE ULTIMATE LEARNING TOOL

SECOND EDITION

Nelson H. Kraus

University of Indianapolis

cognella®
SAN DIEGO

Bassim Hamadeh, CEO and Publisher
Angela Schultz, Senior Field Acquisitions Editor
Carrie Montoya, Manager, Revisions and Author Care
Kaela Martin, Project Editor
Alia Bales, Production Editor
Emely Villavicencio, Senior Graphic Designer
Alexa Lucido, Licensing Manager
Natalie Piccotti, Director of Marketing
Kassie Graves, Vice President of Editorial
Jamie Giganti, Director of Academic Publishing

Printed in the United States of America.

cognella® ACADEMIC PUBLISHING
3970 Sorrento Valley Blvd., Ste. 500, San Diego, CA 92121

To my wife and soulmate, Susan; my children, Erich, Mandy, and Dan; the memory of my parents, John and Alice; my brother, Jack; dozens of university faculty; and the thousands of students who have helped me learn. And finally to my constant helpers, Charlie and Capt. Jack Sparrow!

This text is just like any piece of exercise equipment.
You've got to *use it* to benefit!
So, let's ride!

BRIEF CONTENTS

Acknowledgments xxiii

Preface xxv

Introduction: Building a Foundation of Understanding xxix

FOUNDATIONS: Detailing the Keystones to Understanding Anatomy and Physiology xxxiii

CHAPTER 1 How to Pass A&P with Flying Colors 1

CHAPTER 2 Helpful Flowcharts H_2O Sheet and POSS 13

CHAPTER 3 Body Systems and the Requirements for Life 23

UNIT I Overview of the Human Body 29

CHAPTER 4 A&P Basics: Nomenclature, Terminology, Planes, Body Cavities, and Regions 31

CHAPTER 5 Chemistry Basics for Human Physiology 41

CHAPTER 6 Energy Production and Protein Synthesis 55

CHAPTER 7 Simplified Cell Biology: Organelles and Membrane Transport 65

CHAPTER 8 Histology: Tissues, Glands, and Membranes 77

CHAPTER 9 The Skin: Integumentary System 87

UNIT II Support and Movement 97

CHAPTER 10 Skeleton and Bone Physiology: Calcium Homeostasis 99

CHAPTER 11 The Muscular System and Muscle Contraction Physiology 131

UNIT III **Communication and Control of the Body** **151**

CHAPTER 12 The Nervous System: Structure, Function, and Special Senses 153

CHAPTER 13 The Endocrine System: Glands, Hormones, and Homeostasis 181

UNIT IV **Circulation, Regulation and Reproduction** **189**

CHAPTER 14 The Cardiovascular System: Structure and Function 191

CHAPTER 15 The Lymphatic and Immune Systems: Protection or Defense 209

CHAPTER 16 The Respiratory System: Gas Exchange 215

CHAPTER 17 The Digestive System: Nutrition and Nutrient Absorption 223

CHAPTER 18 The Urinary System: Waste and Balance 233

CHAPTER 19 Nutrition and Metabolism 241

CHAPTER 20 Fluids and Electrolytes: Acid/Base Balance 245

CHAPTER 21 Reproductive System: Continuation of the Human Race 253

Epilogue 259
Glossary 261
Index 269

CONTENTS

Acknowledgments xxiii
Preface xxv
Introduction: Building a Foundation of Understanding xxix

FOUNDATIONS: **Detailing the Keystones to Understanding Anatomy and Physiology** xxxiii

CHAPTER ONE

How to Pass A&P with Flying Colors 1

Excavating for Our A&P Foundation 1
Ten Keys to Learning 2
Medical Terminology 5
Super Simple Basics of Medical Terminology 6
Concepts to Make Your Learning Easier and Your Grade Higher 6
Review Questions 10
Activities 10
Drawing/Coloring Activity 11

CHAPTER TWO

Helpful Flowcharts: H_2O Sheet and POSS 13

How to Effectively Utilize the "H_2O" Sheet 14
Column One—Organizational Hierarchy (OH) 14
Column Two—Macromolecules or Biochemicals 15
Column Three—Primary Tissues 16
Column Four—Systems 17
Column Five—Pathophysiology 19
Physiology on a Single Sheet (POSS) 20
Review Questions 21
Activities 22

CHAPTER THREE

Body Systems and the Requirements for Life 23

Definitions 23
Body Systems 23
Homeostasis 25
Requirements for Life 26
Survival Needs 27
Review Questions 27
Activities 27

UNIT I Overview of the Human Body 29

CHAPTER FOUR

A&P Basics: Nomenclature, Terminology, Planes, Body Cavities, and Regions 31

Anatomical Terminology and Nomenclature 31
Terms 31
Abdominal Quadrants 32
Abdominal Regions 32
Anatomical Planes 33
Open and Closed Body Cavities 34
Serous Membranes and Mesentery 35
Help from Crash Course 36
Anatomical Regions 37
Anatomical Planes 37
Anatomical Position 38
Review Questions 40
Activities 40

CHAPTER FIVE

Chemistry Basics for Human Physiology 41

Chemistry Necessary to Understand Physiology 41
Atomic Structure 42
Major and Minor Chemical Elements Essential in Physiology 42
Chemical Elements Are Displayed in the Periodic Table 42
Isotopes 44

Chemical Bonding 44
Water (H_2O) 46
Dehydration Synthesis and Hydrolysis Reactions 48
Acids and Bases—pH Scale 49
Biological Macromolecules or Biochemicals 49
Enzymes 51
Oxidation/Reduction Reactions 52
Review Questions 52
Activities 53
Coloring Activities 53

CHAPTER SIX

Energy Production and Protein Synthesis 55

Physiology on a Single Sheet (POSS) 55
Cellular Respiration or Metabolism (There's a Difference!) 57
Protein Synthesis (PS) 62
Review Questions 64
Activities and Coloring! 64

CHAPTER SEVEN

Simplified Cell Biology: Organelles and Membrane Transport 65

Cell Biology 65
Typical Human Cellular Structure 65
SPPLBL—Cell Membrane 67
Understanding Cellular Membrane Transport 70
Diffusion/Simple Diffusion 70
Diffusion/Facilitated Diffusion 71
Osmosis 72
Osmotic Pressure and Tonicity 73
Active Transport (AT) 74
Vesicular Transport 75
Review Questions 75
Activities 76

CHAPTER EIGHT

Histology: Tissues, Glands, and Membranes 77

Histology 77
Primary Tissues 77
Epithelial Tissue Types 79
Connective Tissue Types 81
Muscle Tissue Types 83
Nervous Tissue 83
Time to Churn to Learn 84
Review Questions 84
Activities and Coloring 85

CHAPTER NINE

The Skin: Integumentary System 87

Skin or Integumentary System 87
Layers 87
Layers of the Epidermis 89
Layers of the Dermis 89
Review Questions 93
Activities and Coloring 93

UNIT II Support and Movement 97

CHAPTER TEN

Skeleton and Bone Physiology: Calcium Homeostasis 99

Bones 99
Bone Classification 100
Anatomy of Long Bones 101
Skeleton 103
Bone Markings 104
Axial Skeleton—Skull 106
Axial Skeleton—Vertebral Column 108
Axial Skeleton—Thoracic Cage 112
Appendicular Skeleton—Upper Extremity 113

Appendicular Skeleton—Pelvis 118
Appendicular Skeleton—Lower Extremity 119
Joints—Articulations 121
Bone Physiology and the Homeostasis of Calcium 126
Review Questions 128
Activities 129

CHAPTER ELEVEN
The Muscular System and Muscle Contraction Physiology 131

Skeletal Muscles 131
Organizational Patterns of Skeletal Muscle Fibers 132
Naming of Skeletal Muscles 133
Muscle Naming 133
Types of Skeletal Muscles 135
Skeletal Muscle Groups 135
Muscle Compartments 137
External Ocular Muscles 138
Muscle Physiology and the Sliding Filament "Theory" of Contraction 139
Review Questions 148
Activities and Coloring (Draw It to Know It) 148

UNIT III Communication and Control of the Body 151

CHAPTER TWELVE
The Nervous System: Structure, Function, and Special Senses 153

Nervous System 153
Organization of the NS 154
Neuron Structure 154
Structural Classification of Neurons 155
Functional Classification of Neurons 155
Neuroglial Cells—Neuron Support Cells 156
Central Nervous System (CNS) 157
Brain Structure 158
Anatomical Areas of Emphasis of the Brain 158

Midsagittal Section of the Brain 159
Upper and Lower Motor Neurons 162
Spinal Cord and Spinal Nerves 162
Reflex Arc 165
Special Senses—Eye and Ear 165
Nervous System Physiology—Electrical Communication 168
Sensory Nervous System and Special Senses—Eye and Ear 172
Categories of Sensory Receptors 172
Eye Physiology 173
Changing Light into Action Potentials 174
Ion Flux Produces AP 175
Ear Physiology 175
Keeping the Body Upright 176
Review Questions 177
Activities 178

CHAPTER THIRTEEN

The Endocrine System: Glands, Hormones, and Homeostasis 181

Super Simple Summary of Endocrine Glands and Areas 181
Types of Endocrine Control 182
Negative Feedback Loops and Glucose 183
Types of Endocrine Hormones 185
Review Questions 186
Activities 187

UNIT IV Circulation, Regulation and Reproduction 189

CHAPTER FOURTEEN

The Cardiovascular System: Structure and Function 191

CVS = Cardiovascular System: The Body's Transportation System 191
Essential Heart Structures to Know 194
Lymphatic System 199
Cardiovascular System Physiology 200
Review Questions 206
Activities and Coloring 206
Fill in the Indicated Labels on this Diagram of the Heart 207

CHAPTER FIFTEEN

The Lymphatic and Immune Systems: Protection or Defense 209

Innate Immunity 209
Adaptive Immunity 210
T-Cell and B-Cell Lymphocytes 213
Primary and Secondary Response in Humoral Immunity 213
Review Questions 214
Activities and Coloring 214

CHAPTER SIXTEEN

The Respiratory System: Gas Exchange 215

Respiratory System Provides Gas Exchange 215
Muscles of Respiration 217
Ventilation 218
Overview of Respiration 218
Respiratory Volumes and Pressures 219
Gas Exchange 220
Oxygen and Carbon Dioxide Transport 220
Review Questions 221
Activities and Coloring 221

CHAPTER SEVENTEEN

The Digestive System: Nutrition and Nutrient Absorption 223

Digestive System 223
Sphincters 224
Divisions of the DS 227
Nutrient Breakdown 227
Phases of Swallowing 228
Gastric Secretions 228
Regulation of Digestion in the Stomach 229
Digestion and Absorption 229
Review Questions 230
Activities 231

CHAPTER EIGHTEEN

The Urinary System: Waste and Balance 233

Urinary System 233
Filtration 236
Reabsorption 237
Secretion 239
Excretion 239
Review Questions 239
Activities and Coloring 240

CHAPTER NINETEEN

Nutrition and Metabolism 241

Nutrition 241
Lipid Transport 243
Review Questions 244
Activities 244

CHAPTER TWENTY

Fluids and Electrolytes: Acid/Base Balance 245

Fluid Compartments 245
Acid/Base Balance and Buffering Systems 248
Acid/Base Imbalances 250
Review Questions 251
Activities and Coloring 251

CHAPTER TWENTY ONE

Reproductive System: Continuation of the Human Race 253

Reproductive System 253
Similarities of Female and Male Reproductive Systems 255
Male Reproductive Function 256
Female Reproductive Function 256

Review Questions 257
Activities and Coloring 257

Epilogue 259
Glossary 261
Index 269

ACKNOWLEDGMENTS

Many people, both faculty and students, reviewed all or parts of this text, and I am indebted to them all. Student feedback and suggestions were invaluable and extremely helpful.

The following faculty and graduate students at the University of Indianapolis reviewed parts of the text, and their perspectives and guidance was significant:

John Langdon—Biology faculty

Dean Wiseman—Biology faculty

Molly Hill—Biology faculty

Jacob Wood—Biology graduate student

Sammi Beck—Biology graduate student

Barb Kelly—Nursing faculty

Jessica Rivosecchi-Fulton—Biology graduate student

Shawn Hurst—Biology faculty

Nathan Eckert—Exercise Science faculty

Alba Craig—Biology graduate student

Sara Mire—Biology graduate student

Joe Burnell—Chemistry faculty

Arden Mower—Biology graduate student

I also want to thank:

Beda E. Alvarez, Jr., M.D., University of South Carolina Beaufort

Caroline Doyle Ph.D., Bellarmine University

Julia M. Schmitz, Ph.D., Piedmont College

Paul J. Kiser, Ph.D., Bellarmine University

Rollin Medcalf, M.D., Pima Community College

Hai Trinh—City College—San Diego Community College

Finally, a huge thank you to all the folks at Cognella Academic Publishing, especially my acquisition editor and project editor for the second edition, Carrie Montoya and Kaela Martin. Their guidance and patience were extraordinary!

"If you can't explain it simply,
you don't understand it well enough."

—*Albert Einstein*

PREFACE

"OMG, our text is HUGE; what do I really need to know?"

"I don't know what to study!"

"There's so much information, I'm overwhelmed!"

"I'm so confused and stressed!"

"Our text is intimidating!"

"What's on the test?"

"How can I get an A?"

SUPER Simple Anatomy and Physiology (*SS A&P*) overcomes these common questions and concerns from novice learners of anatomy and physiology (A&P). It provides the guidance to build a solid foundation of A&P understanding. It effectively allows students to integrate A&P so they can more easily progress in their education and careers.

SS A&P is unparalleled when utilized as a learning tool text in introductory anatomy and physiology courses. But it is effective in many other situations. Here are some of the areas where *SUPER Simple A&P* excels:

- ✔ Introductory A&P courses (high school and college)
- ✔ Reviewing and refreshing when taking upper-level courses like pathophysiology
- ✔ Reviewing for entrance or certification exams like NCLEX or MCAT
- ✔ EMT/paramedic training
- ✔ Hospital training programs for surgical assistants and surgical technicians
- ✔ Military basic and advanced medic training
- ✔ Review tool for practicing healthcare practitioners like MDs, DPTs, and RNs.
- ✔ Effective patient education tool
- ✔ Fitness centers
- ✔ Massage therapy schools
- ✔ The general public wanting or needing to educate themselves about the human body

Anatomy and physiology are complementary (form follows function, abbreviated as FFF) and inseparable, and many schools teach A&P together. However, some institutions teach each as a separate course. *SS A&P* is structured in a unique manner, allowing it to be easily utilized whether the courses are offered separately or together. Most importantly, the text will be easy for students to utilize as their primary learning tool in any introductory A&P course.

This learning tool is divided as follows:

- Foundations—Detailing the keystones to understanding anatomy and physiology
- Unit I—Overview of the Human Body
- Unit II—Support and Movement
- Unit III—Communication and Control of the Body
- Unit IV—Circulation, Regulations, and Reproduction

The foundations section that is color-coded sea-foam is essential to read and digest (reread, research, and repeat) before diving into the following chapters. Foundations is a "user manual" describing how to get maximum return on investment (ROI) from *SS A&P*. So, take the time and effort to build a foundation of understanding by chewing through the "Foundations" section.

Some chapters, such as chemistry and energy production, only apply to physiology, whereas other chapters, like cell biology and histology or skin, apply primarily to anatomy. Those chapters are color-coded for either anatomy (orange) or physiology (blue). However, most of the chapters throughout *SS A&P* deal with both anatomy and physiology.

Each of those chapters starts with an anatomy section (orange color-coded) followed by a physiology section (blue color-coded). Because A&P are complimentary and inseparable, these two sections are integrated with each other within each chapter and/or to other chapters. This linkage is accomplished by "bridging" between the various chapters and units, as indicated by a bridge icon.

The bridge icon indicates it's important to link to another section, page, or figure of *SS A&P* to clarify concepts and deepen understanding. These bridges make it easier to gain insight by seamlessly blending A&P together. This format also allows the text to be used effectively regardless of A&P course format or sequencing.

Throughout the text, you will see icons that highlight specific ideas or concepts, or direct you to other sections of the book that will deepen your understanding of the text.

TABLE 0.1 Icons found throughout the book

LOOK FOR THE ICON	ICON MEANING	ICON ACTION
	Topics of Emphasis (TOEs) for chapter	Indicates what's coming up
H_2O	H_2O Flow Chart (page 12)	Reminds you to refer to helpful learning tool chart showing A&P on single page
✓ CONCEPT	Highlighting an essential concept to understand	Deepen understanding of basic A&P concepts
	Form Follows Function	Apply concept of form follows function to specific structures
	Move or Bridge to deepen insight	Refer to another section in the same chapter or to another chapter for clarification and better understanding

SS A&P is highly effective in a wide variety of learning environments. However, whatever the learning situation, this guide will minimize memorization while increasing understanding and retention.

SS A&P really does make learning fun, easy, and effective!

ACTIVE LEARNING

This book has interactive activities available to complement your reading.

Your instructor may have customized the selection of activities available for your unique course. Please check with your professor to verify whether your class will access this content through the Cognella Active Learning portal (http://active.cognella.com) or through your home learning management system.

INTRODUCTION

BUILDING A FOUNDATION OF UNDERSTANDING

Typical anatomy and physiology texts cover topics, concepts, and keywords from A to Z, literally, from the anterior pituitary to the zygomatic bone—and everything in between. These excellent reference texts start with the basics of anatomy and physiology (hereafter referred to as A&P), such as the scientific method, anatomical nomenclature, directional terms, and so on, and proceed in detail through chemistry, cellular respiration, cell biology, histology, discussion of all 12 body systems, fluids and electrolytes, nutrition, development, and finally, heredity.

These textbooks are exceptional resources, but they are much more detailed than is needed for the entry-level, novice learner or for any healthcare professional needing to review basic A&P.

SS A&P eliminates a large amount of unessential content so there will be more time to gain a solid foundational understanding of A&P. Building that foundation is essential for a variety of health careers, especially those involving direct patient contact, such as nursing, occupational therapy, physical therapy, exercise science, athletic training, medicine, and others.

It is unrealistic for a novice student to gain detailed insight into the complexity of A&P in only one or two semesters. An introductory course in A&P is only a *first step* in helping a student evolve from novice, with little or no clinical insight, to competent and confident clinical practitioner.

At this introductory level, learning is much like building a four-story building. First, we clear land and dig an excavation. Once that excavation is complete, we'll be framing up to lay footers, as well as driving some pylons if needed for added stability for the upper stories of the building. Then we'll pour concrete for the footers and give those time to set up. That's as far as we'll be going in *SS A&P*. We may get to lay a few blocks for the basement walls, but nothing beyond that. *SS A&P* is all about building a solid foundation. We're basement builders!

SS A&P is a highly effective tool in this metaphorical basement-building process. We're not concerned with the construction of the upper floors or even the first floor. We're constructing a solid foundation, so the first, second, third, and fourth floors will be stable when they are built. This will require persistence, struggle, and repetition. But without the foundation *SS A&P* helps build, those upper floors become irrelevant—and unstable.

A basement is called a foundation because it's essential to the stability of everything built on top of it. *SS A&P* ensures that foundation is rock solid!

A Quick Look at Effectively Utilizing *SS A&P*

To benefit from *SS A&P*, the reader needs to "digest" (read, reread, highlight, underline, write notes in the book, discuss while gaining insight, etc.) the preface, introduction, and the first three chapters. These deal with the process of learning, the concept of form following function (FFF), the H_2O Sheet, and Physiology on a Single Sheet (POSS) flow charts, as well as the basics of medical terminology.

Students may be tempted to just jump into the "meat" of the material as they would with the typical A&P reference text. **Don't do that!** Take the time to read and digest the first thirty pages of *SS A&P*. Doing that will save the student hours of time during the semester/quarter and will help guarantee student success in A&P.

Notice *SS A&P* is concise, light, and printed on non-glossy paper. It's easy to carry around and designed so that students can write notes directly in the text. It is a powerful tool to use preparing for class, during class discussions, laboratory activities, and exam preparation. Although *SS A&P* is color-coded, some students find it helpful to utilize stick-on labels or sticky bookmarks, making it easier to find specific information in *SS A&P*.

Take advantage of the various icons (e.g., FFF, concepts, bridges, etc.) in *SS A&P*. These help the reader integrate anatomy and physiology and take advantage of the concept of FFF while deepening their understanding of A&P.

Follow the guidance to go online and utilize specific resources like Crash Course and Draw It to Know It. Googling topics can also provide opportunities to "view" A&P concepts and details from different perspectives. For example, Googling "Animation of the Sliding Filament Theory of Muscle Contraction" brings up dozens of helpful sites. Find one that allows you to deepen your understanding by watching a fresh perspective. This will augment the basics you find in *SS A&P*.

But, *be cautious* when you are using online sites. Be aware of the legitimacy of a site's information by comparing with *SS A&P* or other resources you know are valid, including your instructor or teaching assistant if you are in a formal course setting.

Finally, utilize the underlying concept of "Form Follows Function" (FFF) to minimize the need for rote memorization. *SS A&P* is designed to create A&P understanding and insight. When you understand, you don't need to memorize as much, and you'll remember more, which will benefit you in coming courses and your future healthcare career.

Form Follows Function

Knowing the function of an organ or structure provides some insight into how it may be structured. So, form follows function, but *structure will reveal* function.

For example, the heart is shaped the way it is because of what it does, functioning as a unified and synchronized whole. It is imperative for the heart to function as a syncytium (a great Scrabble word meaning a mass of cells that have merged together). The heart must always be NSYNC! And because the heart is shaped to perform a very specific function, its shape or structure tell us something about its function. Function determines structure. Structure reveals function.

This guide will constantly push the student to integrate anatomy and physiology by applying FFF. There will be FFF icons throughout the text highlighting the complementary nature of anatomy and physiology.

Most importantly, remember, it is imperative for the student to put in the effort to learn and understand A&P. This is true when learning anything, playing the piano, tennis, golf, a foreign language. The one doing the learning has to put in the effort or they don't learn much, if anything.

In any learning environment, the ones doing the most work are the ones doing the most learning. So, I encourage you to be one of those putting in the most work!

DETAILING THE KEYSTONES TO UNDERSTANDING ANATOMY AND PHYSIOLOGY

CHAPTER ONE

How to Pass A&P with Flying Colors

TOES (TOPICS OF EMPHASIS)

- What you must know to be successful in A&P
- Ten learning keys for success in A&P
- Medical terminology basics
- Concepts that make your learning easier and your grade higher

Excavating for Our A&P Foundation

Let's get started building the solid foundation of A&P understanding by excavating for the basement of our four-story building described in the introduction. The primary basic concept of anatomy and physiology is *"FFF"* or *form follows function*. A&P are inseparable and complementary. They are interrelated, and understanding the concept of FFF dramatically improves insight and understanding of A&P.

Anatomy is the study of human structure (form), and physiology is the study of human function. These two sciences are inseparably linked together. *Organs and organ systems in the body are structured the way they are because of what they do.*

In other words, function (physiology) determines structure (anatomy). An understanding of one of these sciences is helpful, arguably essential, to understand the other science. Because both anatomy and physiology are sciences, it is important for the novice A&P student to have a basic understanding of the scientific method (SM).

The *scientific method* (*SM*) is the standard process of observation, experimentation, and data collection that has allowed the development and evolution of modern-day sciences, like A&P. It goes all the way back to Aristotle and the Greeks. The *SM* allows scientists to confirm theories and build on existing knowledge. Many of the concepts discussed in physiology are theories developed using the scientific method.

The SM starts with simple observations and subsequent questions. An example would be "how do cells make energy?" A hypothesis or explanation to the question is developed. Then an experiment is designed to test that hypothesis. Data is collected from the experiment and analyzed. Scientists determine if the data supports the hypothesis. Depending on the conclusions drawn from the experiment, scientists either design additional experiments to support the hypothesis or develop a modified hypothesis. If the data supports the hypothesis it will then be further tested by other scientists in distant locations. If the data from multiple independent experiments confirms the hypothesis, it will become a theory answering the original question. That's how scientists developed the sliding filament theory of muscle contraction or the sodium-potassium pump theory, which maintains the resting membrane potential (RMP) in a neuron. Don't worry about understanding those theories right now; we'll get to those later in *SS A&P*, although you can bridge to those sections if you're eager to learn more right now! ;o)

BRIDGE to pages 141–144 and 169 for details

The steps of the *SM* can be summarized as follows:

- Observation
- Question
- Hypothesis
- Experimentation
- Data analysis
- Conclusion

The scientific method is the "gold standard" that differentiates true scientific conclusions from just opinions and pseudoscience. The basic concepts of A&P have been derived using the scientific method. That's how we know the cell makes energy through cellular respiration (CR), which involves four specific stages. We'll discuss details of CR in Unit IV of *SS A&P*.

Bridge to chapter 6 for a detailed discussion of CR or search the internet to learn more.

BRIDGE to Chapter 6 for more about CR

Ten Keys to Learning

Understanding how the human learns will help students understand and benefit from the activities in *SS A&P*. The author has identified ten essential building blocks that facilitate human learning.

1. The first *requirement* for effective human learning is **repetition**.

It is the first key necessary to learn anything. Learning a language, playing a musical instrument, or learning how to hit a golf ball all require a ton of repetition or practice. Repetition is an essential key and *SS A&P* provides an opportunity to do constant repetition. Content that is not necessary at an introductory level has been eliminated, so you can more easily practice repetition. With this approach we'll "churn to learn" (CTL)! Because it's easier for most people to remember a picture than words, Figure 1.1 shows a person "churning" butter and also reading. She's "churning to learn," just like you will throughout *SS A&P*.

FIG. 1.1 Churn to Learn (CTL)

2. **Emotions** are the second key to human learning.

These can be either positive or negative emotions unless they're carried to an extreme. Of course, it is better to be happy in a learning experience, but there will be times of frustration, unhappiness, and struggle. Struggling is an important part of the learning process. In any learning endeavor, *avoid boredom by creating an environment that offers opportunities for emotions as well as repetition.* SS A&P *will help create an environment of "emotional repetition."*

3. A third key to learning is **voicing**, vocalizing, or just plain talking.

Talk to other students. Read out loud! Explain things out loud. Talk out loud to yourself! *Struggle to teach* someone else an aspect of A&P. You'll *always* learn more when teaching than you will as a student. So, *teach to learn.*

4. **Visualization** is a fourth key to learning.

It is easier for most of us to remember a picture than a page full of words. The old saying "A picture is worth a thousand words" is absolutely on target. This is particularly true for a subject like gross anatomy (study of body parts you can see with the naked eye) and histology (the study of tissues). The flow charts (e.g., H_2O Sheet, POSS [physiology on a single sheet], etc.), which are found in *SS A&P,* are designed to help students get a picture in their mind's eye of some of the essential concepts of A&P. Strive to structure concepts as visualizations or pictures. Draw diagrams. Color pictures. Utilize an anatomy coloring book. Visualize! Work collaboratively with other students to structure Google Slides or Prezis of A&P topics. Draw It To Know It.

5. A fifth essential key to facilitate learning is to **focus on concepts**.

Most A&P courses dump an overwhelming amount of content on students. Students are forced into memorization and regurgitation without real understanding. They do this to survive and obtain the grade needed to move on in their academic careers. But this mindless memorization and regurgitation (pump, dump, and forget) doesn't allow students to retain much, if any, anatomy and physiology. As a result, students are forced to relearn A&P in future courses, clinical rotations, or on the job after graduation. *SS A&P* helps students gain insight into anatomy and physiology, thus retaining more because they really understand. This allows students to avoid most of the need for pump and dump. After all, a basic understanding of A&P is essential to allow medical personnel to deliver high-quality care.

6. A sixth key to learning is **communication**.

This compliments the third key of vocalization. Talk with other students. Attempt to explain concepts to a roommate, family member, or friend. Communication will be an essential skill in any area of healthcare.

7. **Collaboration** is a seventh key to learning.

This compliments several of the other keys of learning like vocalization and communication. Collaborate and cooperate with other learners. Everyone benefits when you help others gain insight into the material. *Teach to learn.*

8. An eighth key to learning is **association**.

It is effective to relate new information or concepts to existing knowledge. *SS A&P* takes advantage of this learning key. For example, we will relate the circulatory system (CVS), which is the body's transportation system, to transportation systems in our communities. This involves vehicles, an engine providing energy, and highways where the vehicles travel. That relationship makes understanding the CVS easier and more fun.

9. The ninth key to learning is movement.

A growing amount of scientific evidence is showing movement and cognition are powerfully connected. Even something as simple as taking a walk while talking with a study buddy about A&P (or any other topic) helps learning. So, get off your gluteus maximus and stretch your lower extremities on a regular basis while churning to learn!

10. The tenth key to learning is memorization.

Memorization is a part of the process of learning anything. But, by emphasizing basic concepts, students are more likely to develop a real *understanding* of anatomy and physiology. This, of course, will require some memorization, but gaining real insight into A&P minimizes memorization, improves retention, and results in higher exam scores, better grades, easier admission into professional programs, and greater success in a healthcare profession.

Let's do some repetition and review the ten keys to learning:

REPETITION—the most essential key that is also important for memorization

EMOTIONS—either positive or negative

VOCALIZING—talking about a subject (strive to teach it to someone else)

VISUALIZATION—"seeing" images of structures, processes, and concepts in the "mind's eye"

CONCEPTS— grasping concepts facilitates and simplifies understanding, especially at an introductory level

COMMUNICATION—talking to others, discussing concepts while you struggle to learn

COLLABORATION—helping other people learn; the more you help others, the more you'll learn

ASSOCIATION—relating new knowledge to existing knowledge

MOVEMENT—a moving body learns more effectively

MEMORIZATION—always a part of learning anything but NOT learning by itself!

Remember these keys with the acronym REVV CCCAMM: rev cam, like a revolving camera!

And because it's easier to remember pictures than words, or ten letters strung together, we can apply illogical or absurd visual association. Visualize REVV CCCAMM by picturing people using their smartphone cameras while rotating around on a sky carousel shooting video, as shown in Figure 1.2. Silly, but it is easier to remember revolving camera, a "rev cam" than it is a series of words.

REV CAM - Revolving Camera

FIG. 1.2 Spinning Sky Carousel Rotating, Camera Aperture—REV CAM Revolving Camera

NOTES

Medical Terminology

Understanding medical terminology makes learning A&P easier and may be a prerequisite for some A&P courses. *SS A&P* is not designed to provide a comprehensive overview of medical terminology, although medical terms are utilized throughout, as they are in any A&P text. Although not absolutely needed to build your foundation of A&P insight, it is helpful to learn or review medical terminology when studying anatomy and physiology.

There are online resources for learning medical terminology. Many of these utilize the memory techniques of absurd or illogical visual association and/or "audionyms" to facilitate efficient understanding and retention. Audionyms are sound-alikes. For example, "gastro" which means stomach, is easier to remember if we apply an audionym. Gastro sounds like gas truck. Visualize a gas truck with a big tank (full of gasoline) on the back. Now, replace that tank (in your mind's eye) with a huge human stomach, flesh colored with arteries and veins on the surface and anchovies spilling out the top of the stomach opening to the esophagus. Totally illogical but an unforgettable image, thus easy to remember. It's silly but it works! We'll be utilizing those same techniques in *SS A&P* to make it easier, quicker, and more fun when learning A&P!

Remember, becoming fluent in medical terminology is *essential* for anyone going into any health-care profession, so you might as well become fluent early.

Even though *SS A&P* is focused on conceptual understanding, learning to more effectively utilize your memory is invaluable! *Let's face it, memorization is an important part of learning anything!*

Another useful tool we will reference is "Draw It to Know It" (Ditki). This online site provides an effective means for students of all medical specialties, MD, DO, allied health, dentistry, nursing, and more. Drawing structures, pathways, and processes harnesses the kinesthetic power of active learning! The *SUPER Simple Lab Manual* utilizes this kinesthetic power with drawing and coloring activities for the A&P laboratory. *Ditki* is a good complement to the *SS A&P* text and lab manual and will benefit the student who is focused on getting a top grade while gaining a real understanding of A&P.

Super Simple Basics of Medical Terminology

If you don't want to spend extra time learning some medical terminology right now, you will benefit by at least learning the mere basics of medical terms and how they are structured. Even this limited insight will help you as you move forward learning anatomy and physiology.

Medical terms have three possible components. Understanding these components and how they combine will make it easier to learn A&P and will be invaluable in any healthcare profession.

Root—This component reveals the primary meaning of the medical term.

Prefix—This is the beginning part of a medical word and helps define the word root meaning.

Suffix—This ending of the word further defines or changes the overall meaning of the term.

Medical words are formed by combining the word root with a combination of another word root, prefix, or suffix. There are comprehensive listings of roots, prefixes, and suffixes online in a variety of sites. Plug "medical word roots" or "prefixes" or "suffixes" into any browser to find those lists in sites like Wikipedia.

An example of how medical words are formed is endocarditis. "Endo" is a prefix meaning "inner." The word root "card" means heart and the suffix "itis" means inflammation. So, endocarditis means "inflammation of the lining of the heart." Most medical terms are structured in this same pattern. Some medical terms will utilize a combining vowel, an "i" or "o", between components that makes the word easier to pronounce.

You will see word roots, prefixes, and suffixes throughout *SS A&P*, so the more insight you have into medical terminology, the easier anatomy and physiology and then pathophysiology will be for you.

Concepts to Make Your Learning Easier and Your Grade Higher

FFF

The concept of form follows function or FFF is a primary underlying theme throughout *SS A&P*. FFF originated in architecture. What a building is used for determines how it's structured. Structure follows function. FFF was first coined by the American architect Louis Sullivan in his article "The Tall Office Building Artistically Considered" in 1896.

FFF can also be applied to the design of cars, trucks, military vehicles, and so on. The cement truck in Figure 1.3 is a good example of FFF. It is structured the way it is because of what it DOES! Function determines form. Physiology determines anatomy. *Conceptually, how an organ is structured is determined by what it does! Form follows function! This concept will be constantly integrated into this text to help facilitate student understanding!*

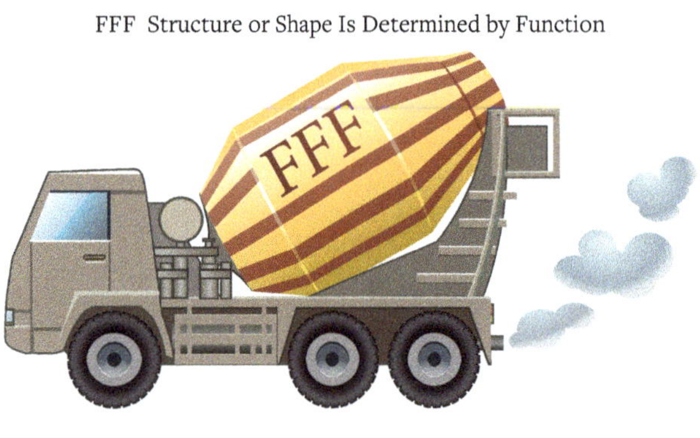

FIG. 1.3 Function of a cement truck determines the structure of the truck

NOTES

How human organs and organ systems are structured is determined by what that structure does! As we're doing right now, we'll revisit FFF frequently. Once a student gets her or his head "around" the concept of FFF, the more likely that student will be successful building a foundation of basic understanding of A&P and improving outcomes on exams, as well as increasing retention.

Because form follows function, the structure of an organ will *reveal* something about what that organ does. For example, the lungs' conceptual function is gas exchange, oxygen moving out of the lungs and into the blood stream and carbon dioxide out of the blood stream and into the lungs. This process takes place by simple diffusion, a passive process in gases and liquids where substances move from higher to lower concentration. To facilitate this process the walls of the gas-exchanging sacs of the lungs (alveoli) are very thin. This is a great example of FFF.

Gas Exchange in the Lungs

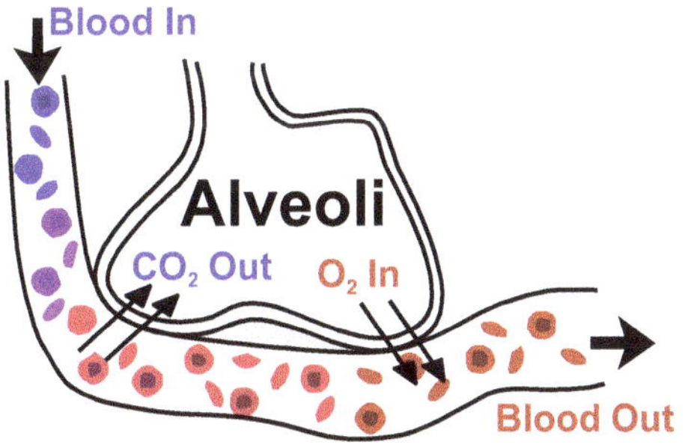

FIG. 1.4 Thin membrane in alveoli facilitates gas movement in the lungs

If you grasp the concept of FFF and observe the ultrathin lining of the alveolar sacs, you can deduce there must be an exchange of substances across those thin sac walls. On the other hand, when observing the thickness of the skin in the human, it is logical to deduce that skin offers a protective barrier and does not facilitate diffusion of gases or liquids. *SS A&P* will focus on these types of concepts to build a basic understanding of the human body, its structure and function.

Homeostasis

Another fundamental A&P concept is *homeostasis*, which is the body's ability to maintain a relatively constant internal environment despite a changing external environment. Our body maintains temperature, levels of electrolytes, blood glucose, and blood calcium levels through the process of homeostasis. This concept will recur frequently in *SS A&P.* It is worth a mention now because understanding function can help a student learning anatomy anticipate or predict how organs are structured. The body makes perfect sense because things are structured the way they are because of their function (what they do).

Homeostasis can most concisely be defined as "dynamic constancy" or DC. Human physiology can be described in this single word "Homeostasis", which is the body's ability to maintain a relatively constant internal environment. Thus, homeostasis can be referred to as the "BIG H" word.

BRIDGE
to Chapters 10 and 13 for greater insight about homeostasis

Let's look conceptually at the DC of blood glucose levels. Our cells utilize glucose as the primary source of energy production. So, it is essential for our bodies to keep our blood glucose levels within a narrow range. When we lose our ability to maintain homeostasis of glucose or thousands of other substances, we become unhealthy, diseased, or dysfunctional. The homeostatic balance of glucose (sugar) and calcium are two of the most important concepts for students to understand. *SS A&P* will concentrate on developing a solid foundation of understanding the homeostatic processes for both blood glucose and calcium.

Normal blood fasting glucose levels range from 70–100 mg/dl. The body maintains this range through "negative feedback loops." These can be easily understood by comparing them with an HVAC (heating, ventilation, and air conditioning) system in a building. The thermostat is set at 70 degrees. When the temperature drops to 69 degrees, the system turns on the heat to raise the temperature. Once it reaches 71 degrees the heat turns off. This maintains a constant temperature even though there is some fluctuation between 69 and 71 degrees. The HVAC system provides a *dynamic constancy.* Human homeostasis does the same. We'll look at homeostasis in more detail as we move through *SS A&P,* constantly helping build a basic understanding of this vital process.

BRIDGE
to Chapter 13 for details about glucose homeostasis

Human Organization and Homeostasis Sheet

THE HUMAN ORGANISM & HOMEOSTASIS (Remember FFF!)

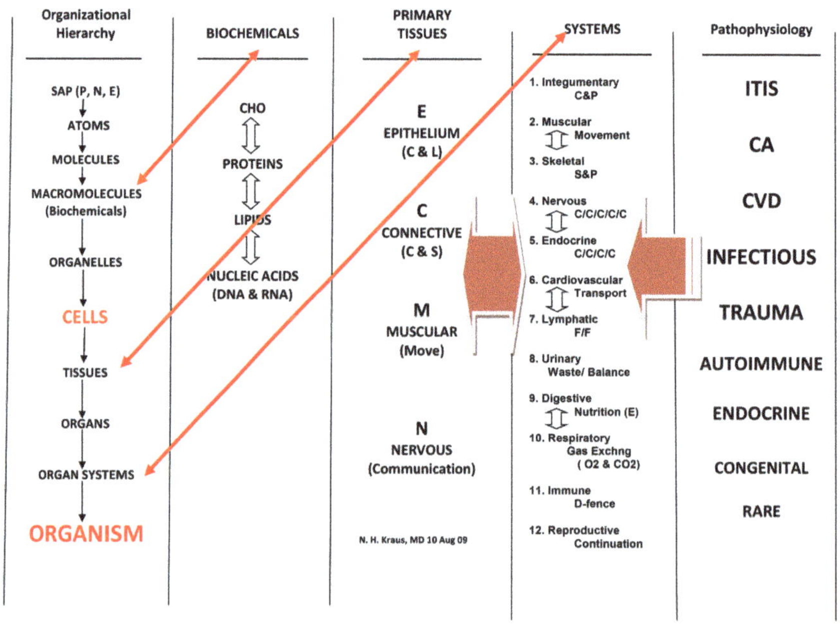

FIG. 1.5 Human Organism and Homeostasis

Flow Charts

Finally, *SS A&P* provides a variety of flow charts to help simplify and conceptualize A&P. These are designed to summarize ideas, minimize memorization, and facilitate insight. Flow charts are one type of learning tool that help students gain insight into A&P.

The "Human Organization and Homeostasis Sheet" (H_2O sheet) is one of these flow charts. We'll be "dissecting" the H_2O sheet in chapter 2 and then utilizing it as we progress through *SS A&P.*

BRIDGE
to Chapter 2 to gain insight into the H_2O Sheet

"Crash Course" with Hank Green is another fun and effective resource to help the novice A&P student "churn to learn." Hank is entertaining and informative. There are dozens of Crash Course lessons, each ten to fifteen minutes long, covering all the topics in A&P. Just put "Crash Course Anatomy and Physiology" into your browser, surf around, find a segment covering an A&P topic you want to review, and *churn* away with Hank. He goes at a fast pace, so it's helpful to stop the video to absorb what is being discussed. I'd also suggest repeat viewings of Crash Course lessons. *Remember REVV CCCAMM*!

"Human Physiology on a Single Sheet" (POSS) is another helpful flow chart. We'll be utilizing POSS as we continue our learning. It will be explained in more detail in chapter 2. If you want to gain more insight into POSS right now, just bridge to chapter 2.

BRIDGE
to pages 18–19 for POSS details

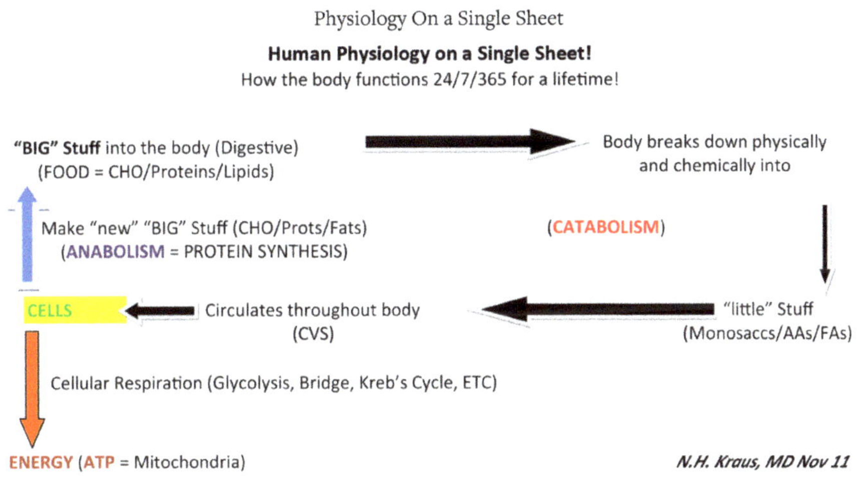

FIG. 1.6 Human function (physiology) on a single super simple sheet

NOTES

Review Questions

(Write the answers on a separate sheet of paper.)

- What is the concept of "FFF"?
- What is the scientific method, and why is it important?
- What are the steps involved in the scientific method?
- What are ten essential factors that facilitate human learning?
- What is the meaning of the term "homeostasis"?
- Why is homeostasis referred to as the big "H" word?
- What is a "negative feedback loop"?

Activities

(Writing these out on a separate sheet of paper can help understanding.)

- Explain why the comparison of DC (dynamic constancy) or homeostasis with an HVAC unit is an effective means of gaining insight into physiology.
- Review and discuss (with another student) the ten *keys* to learning.
- Explain the concept of an audionym.
- Explain the scientific method to a ten-year-old.
- Struggle to write out a list of six *good* examples of FFF in A&P in the human or in architecture, furniture, or vehicles. Explain how your examples demonstrate FFF.

NOTES

Drawing/Coloring Activity

Use the space provided to draw a conceptual graphic demonstrating FFF in the human body. Remember, "draw it to know it" (ditki).

FIGURE CREDITS

Fig. 1.1: Source: https://commons.wikimedia.org/wiki/File:Churn_(PSF).jpg.

Fig. 1.2a: Source: https://pixabay.com/photos/sky-carousel-entertainment-travel-3279551/.

Fig 1.2a: Source: https://pixabay.com/photos/camera-aperture-digital-camera-dslr-1239384/.

Fig. 1.4: Copyright © Helix84 (CC BY-SA 3.0) at https://commons.wikimedia.org/wiki/File:Alveoli.svg.

CHAPTER TWO

Helpful Flowcharts

H_2O Sheet and POSS

TOES (TOPICS OF EMPHASIS)

- Understanding the "Human Organism and Homeostasis" (H_2O) sheet
- Using the H_2O sheet to minimize mindless memorization
- Understanding the POSS and its summary of physiological processes

Human Organization and Homeostasis Sheet

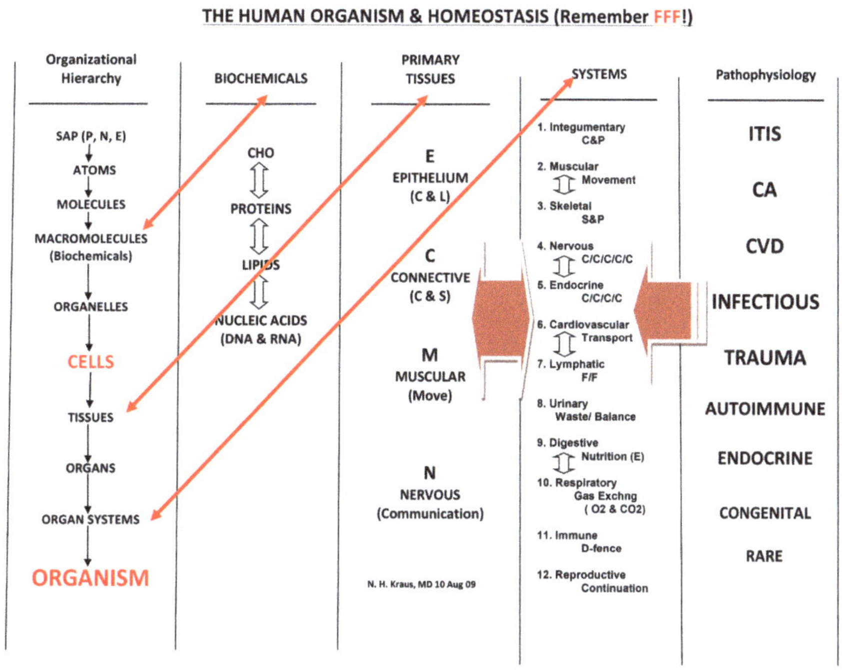

FIG. 2.1 Human Organism and Homeostasis

How to Effectively Utilize the "H_2O" Sheet

The "Human Organization and Homeostasis," or H_2O sheet, was mentioned in chapter 1. The H_2O sheet summarizes A&P, plus pathophysiology, on a single page. In this chapter, you'll learn how to use the H_2O sheet to provide insight into A&P.

H_2O sheet is a valuable resource

Water (H_2O) (two Hs and one O) makes up about two-thirds of the human body mass, which makes the H_2O sheet a little easier to remember. You can visualize a picture of a drop of water, the ocean, rain or a stream to remember the H_2O sheet.

SS A&P uses the simple icon to the left to remind you to refer to the H_2O sheet to help you better understand the topic being discussed. Although memorizing the sheet may be helpful, working to *understand* its structure will make it an invaluable tool.

Let's discuss the flowsheet's layout in detail: It's divided into five columns: organizational hierarchy (OH), biochemicals (macromolecules), primary tissues, body systems, and pathophysiology.

Notice that macromolecules from the first column (OH) is connected by the red line (with arrow heads) to the second column, macromolecules. Likewise, tissues are connected by another red line to the third column, primary tissues. Finally, systems from the first column, OH, is connected to the fourth column listing the 12 systems of the human. Also notice the large, double-headed arrow between the third and fourth columns. This indicates that all 12 of the systems are made up of a blending of the primary tissues (ECMN). Finally, the large red arrow pointing from the fifth, pathophysiology column, to the fourth systems column, indicates all of the systems can be affected by any of the types of pathology, like inflammation or cancer.

Remember, in the human being, *everything* is connected to *everything* else in a variety of ways. We will take advantage of those connections while building our foundational understanding of A&P.

NOTES

Now, let's look at each column in the H_2O sheet.

Column One—Organizational Hierarchy (OH)

The human organism is a complex structure made of very small "parts." Organizational hierarchy (OH) constructs the parts of the body from the most basic to the most complex. Another way to think about this, is that little to big forms human anatomy.

The entire human body is made up of only three *subatomic particles* or SAPs: protons (positive charge), neutrons (neutral charge), and electrons (negative charge). These three SAPs make up every atom on the periodic table of elements.

The most important atoms for human structure and function are carbon, hydrogen, oxygen, and nitrogen. These elements bond together to form molecules and macromolecules or biochemicals. Macromolecules are the basic organic (carbon-based) molecules that make up our bodies. They're also the primary sources of nutrition: comprising carbohydrates, proteins, lipids, and nucleic acids. These organic molecules create cells and the intracellular organelles found inside the cell. Cells join to form tissues, tissues group to form organs, organs form organ systems, and organ systems (humans have twelve) create the human organism!

Figure 2.2 provides a visualization of organizational hierarchy.

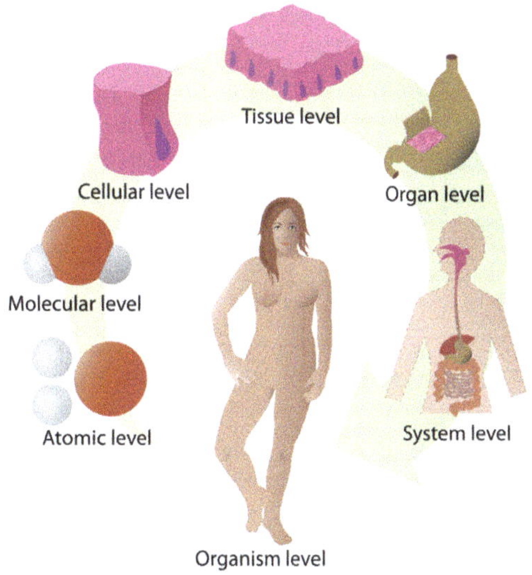

FIG. 2.2 Little to big forms the human body

✔ **CONCEPT**

Concepts are KEY to understanding A&P, Little to big forms the human body.

NOTES

Column Two—Macromolecules or Biochemicals

The four organic (contain carbon) molecules making up the human body are listed in column two of the H_2O sheet. These are CHO or carbohydrates (which contain only carbon, hydrogen, and oxygen), proteins (made from amino acids), lipids (fats), and nucleic acids (DNA and RNA). They are listed in the order in which they are digested: **C, P, L, N.**

Digestion begins in the oral cavity for carbs, the stomach for proteins, and the duodenum for fats. Nucleic acids are found in animal and plant sources of nutrition but aren't substantial enough to be considered a nutrient source. So, we'll think of C, P, and L as the three sources of human nutrition.

Notice the *macromolecules* in column one (OH) are connected to column two with a red arrow. This indicates the integration of the human body. Everything in the body is connected!

NOTES

Column Three—Primary Tissues

Just as the human is primarily composed of only four biochemicals (C, P, L, N), it is made up of only four primary tissues: epithelium, connective, muscle, and nervous (E, C, M, N). Under each of the primary tissues there are two letters or a word indicating their conceptual functions.

Notice in the H_2O sheet, Figure 2.1, epithelial tissue covers and lines as denoted by the two letters (C & L). So, conceptually, any tissue, anywhere in the body, that covers or lines a structure or organ will be made up of one of the types of epithelial tissue. The conceptual function of epithelium may be easier to remember with the visual in Figure 2.3.

If it is tissue covering or lining a structure, then it's made up of *epithelium*! We will discuss as many as nine different types of epithelial tissue when we study histology in chapter 8.

Connective tissue (CT) connects and supports (C & S) the body. So, tissues that connect (tendons, ligaments) and tissues that support (bone, blood, cartilage, etc.) are connective tissues (CT).

CT is the most varied of the four primary tissue types and includes some unusual tissues like adipose (fat) and blood. We'll study 13 specific types of CT in chapter 8.

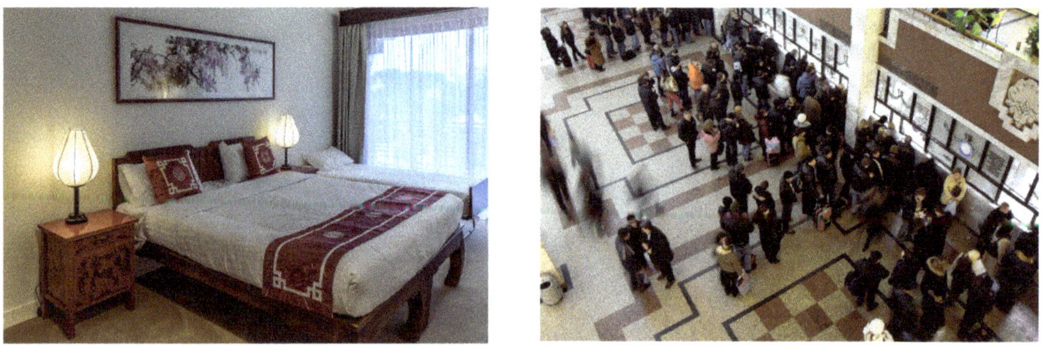

FIG. 2.3a,b Bed Spread and Queue at Ticket Window—Epithelium covers and lines

Muscle tissue creates movement (move) in the body with very few exceptions. Whenever there is movement of, or within, the body, muscles are involved. Muscle tissue can only do one active thing, shorten or contract.

There are three different types of muscle tissue: cardiac, visceral (smooth or involuntary), and skeletal (striated or voluntary). We'll cover those in chapter 8.

Nervous tissue provides communication in the body. Neurons are the only type of cell in nervous tissue that transmit an impulse. There are also six different types of neuroglial cells that support neurons. Different structural and functional types of neurons are discussed in chapter 12.

Notice the large red arrow pointing between columns three and four. This indicates all the body systems are made up of a blending of the four primary tissue types, E, C, M, N!

NOTES

Column Four—Systems

The H_2O sheet lists the 12 organ systems of the human body. Organs (such as the stomach, liver, or heart) are a blending of the four primary tissues (E, C, M, N). Organ systems are groups of organs joined to perform a function.

Notice there are words and/or abbreviations below each of the systems in column four representing their conceptual function.

The **integumentary system** (skin) contains and protects (C&P). Skin holds in the body's fluids and keeps out foreign contaminants and bacteria. In fact, when large amounts of skin are lost, in a severe burn for instance, the body will lose large amounts of fluid, causing life-threatening hypovolemic shock. Because the skin is such an important part of the immune system (defense), skin loss also increases the risk of infection, and once the patient's fluid balance has been stabilized, steps must be taken to protect the wounds from infection.

The **muscular system** (skeletal muscles, such as our biceps brachii) provides movement. The muscular system is made up of skeletal muscle. These muscles appear striped under a microscope and so are also called striated (banded) muscle.

All skeletal muscle movement is voluntary (walking, playing the piano, breathing, speaking, or moving the eyes). You can remember the diaphragm (main breathing muscle) is a skeletal muscle because we can control our breathing.

Skeletal muscles originate from bones and cross joints and connect to other bones. When those skeletal muscles contract, movement of the skeleton occurs, allowing us to walk, run, and dance.

The **skeletal system** (skeleton) is composed of bone and cartilage that supports and protects (S&P) the body. It is sometimes combined with the muscular system to form the musculoskeletal system (MS). However, we're going to separate them to better grasp the individual aspects of each. However, remember they work together to create body movement.

Notice the vertical white arrow connecting these two systems on the H_2O sheet; this is a reminder they work together to move the body.

Most skeletal muscles originate from bones and cross joints and connect to other bones (the platysma muscle and extraocular eye muscles are exceptions). When those skeletal muscles contract, movement occurs. Joints or articulations allow for movement, but they are prone to injury and disease because they carry so much of the stress associated with walking, running, or dancing.

NOTES

The **nervous system** (**NS**) is the body's primary communication system. The NS communicates electrically, instantaneously, and with very short duration of action. It **provides** communication, commands, controls, coordinates, and allows consciousness. These "five Cs" are noted under the NS in the H_2O sheet. We'll discuss the structure of the NS and how it functions in detail in chapter 12.

The **endocrine system** (**ES**) is the body's secondary communication system. It provides all the same functions as the NS with the exception of consciousness. So, it has four Cs instead of five. That's why it's connected to the NS with the vertical white arrow, just like the MS systems.

The ES communicates chemically (with hormones), relatively slowly when compared with the NS, and with long duration of action. We'll cover the ES when we get to chapter 13.

BRIDGE
to CVS Flow Chart on page #191, Fig. 14.1.

The **cardiovascular system** (**CVS**) is the body's transportation system (transport). With few exceptions, when the body needs to move "stuff" (blood, hormones, nutrients, etc.) from one place to another, it uses the CVS to do the moving or transporting.

The CVS is like a subway system: Trains (blood cells) travel through highways or tunnels (blood vessels) and are moved with energy from engines (the heart). When we get to the physiology of the CVS, we'll continue to use this analogy as part of a CVS flowchart.

The **lymphatic system** (**LS**) is primarily a part of the immune system (defense) but also provides a secondary transportation function. It is placed right under the CVS on the H_2O sheet and is connected to it with another vertical white arrow.

The LS absorbs interstitial fluids from between cells as well as digested fat molecules from the small intestine and then filters the lymph composed of those absorbed materials. So, the LS absorbs *fluids* and *filters* lymph (F/F).

The **digestive system** (**DS**) provides nutrition (C, P, L) to the body so the cells can manufacture energy stored in the form of ATP (adenosine triphosphate).

Energy is denoted in the H_2O chart by the letter "E."

The DS has two anatomical divisions, the gastrointestinal tract (GI tract), or the alimentary canal, and the accessory organs, like salivary glands and the pancreas.

The **respiratory system** (**RS**) has a primary function of gas exchange. It allows us to breathe in air and provide oxygen to our cells (needed to efficiently produce ATP) and to remove carbon dioxide, a waste product from energy production.

Once again, there is a white vertical arrow between DS and RS because they work together to efficiently provide the necessary raw materials (nutrients and oxygen) for cellular energy production.

The **urinary system** (**US**) filters waste from the blood and balances "stuff" like glucose, acidity, and electrolytes (sodium, potassium, etc.). The kidney is the center of the system, a marvelous organ providing a vital cleansing function to the body.

The H_2O sheet shows waste/balance as the conceptual function of the US.

The **immune system** (**IS**) is all about *defense*. There are two divisions of the IS, the innate immune system and the adaptive or acquired immune system.

✔ **CONCEPT!** Frequently in A&P one item will be divided into two subdivisions. This is the "rule of two." We'll see this over and over as we move through *SS A&P*.

We are born with the innate immune system consisting of barriers protecting against a wide variety of challenges. Structures like the skin are part of the innate immune system.

Inflammation and special white blood cells like "natural killer" cells also provide innate protection. The adaptive immune system responds to specific, individual challenges by producing antibodies that overcome a specific invader.

The **reproductive system** (**RS**) is the only system not necessary to sustain an individual's life. But, without it none of us would be here, and life certainly wouldn't be as interesting. The RS allows our species to continue after the individual's death (*continuation*).

NOTES

Column Five—Pathophysiology

SS A&P is primarily concerned with normal structure and function, but the fifth column of the H_2O sheet provides a "glimpse" of what can go wrong (dysfunction or disease) with any of the 12 body systems.

Notice the large red arrow pointing from the pathophysiology column back to the systems column. This indicates any of the diseases can affect any of the systems. The fifth column uses abbreviations indicating the most common causes of disease and dysfunction in humans: inflammation (ITIS), cancer (CA), cardiovascular disease (CVD), infectious diseases (infectious), trauma (caused by blunt force, burns, etc.), autoimmune diseases (where the body's immune system attacks its own tissues; e.g., DM (diabetes mellitus) Type 1, rheumatoid arthritis, etc.), endocrine diseases (involving the endocrine system), and finally congenital (born with a dysfunction) and rare diseases (from Aaroskog Scott syndrome to Zuska's disease).

The H_2O sheet is designed to provide an effective summary of anatomy, physiology, and a little pathophysiology on a single page. Spending time "digesting" this sheet in conjunction with the explanations just provided. Will be extremely helpful. Developing a real understanding of the structure and function of the human body will be more fascinating than memorizing isolated bits of information.

NOTES

Physiology on a Single Sheet (POSS)

Taking the time to understand this flowchart will save you hours of struggle later, so let's "dissect" the POSS in detail and churn to learn (understand!) it right now!

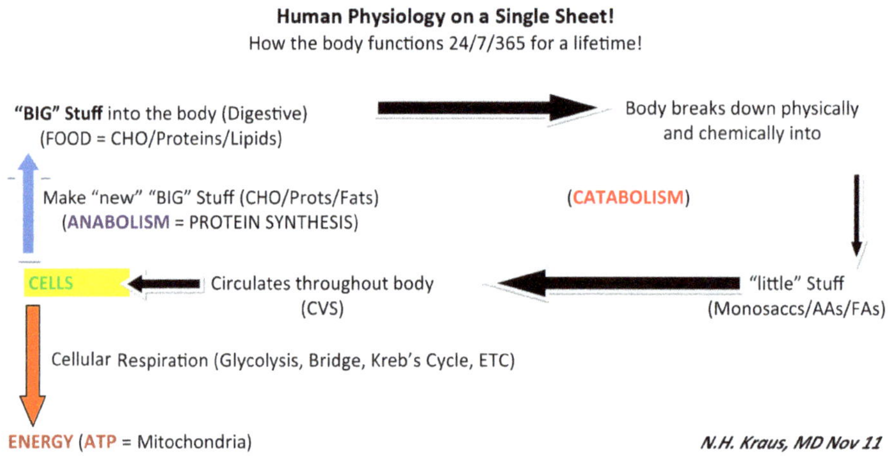

FIG. 2.4 Human function (physiology) on a single super simple sheet

The POSS is a classic example of simplifying complex topics. Presented in the form of a graphic, it is much easier to visualize, understand and utilize to minimize memorization

Look at the upper left-hand portion of the POSS. No matter what else we're doing, our bodies are repeating the same activities needed to maintain human life. We must consume food and drink (water) containing the three basic sources of nutrition: carbs, proteins, and fats (C, P, L) on a regular basis to sustain life.

Our food sources, both animal and plant, also contain nucleic acids (DNA and RNA), but since these occur in such small amounts, they are not a major source of nutrition. So, for the sake of simplicity, we'll ignore the nucleic acids.

The three major sources of nutrition are broken down through both physical digestion (chewing and churning) and chemical digestion (digestive enzymes) in a process called catabolism (the process of breaking down large molecules into smaller molecules)

H_2O **BRIDGE** to the H_2O sheet to review the four primary biochemicals

Carbs are broken down to monosaccharides; proteins into amino acids; and lipids into fatty acids. Those small molecules are then circulated throughout the body by the CVS.

Look at the fourth column on our H_2O sheet and find the cardiovascular system. Underneath the CVS is the word "transport." The CVS is the body's "transportation" system.

✔ **CONCEPT!** With an occasional exception, whenever our bodies need to move "stuff" (nutrients, wastes, hormones, etc.) from one place to another in the body it involves the CVS.

Back to the POSS, the digested monosaccharides, amino acids, and fatty acids from food are circulated throughout the body to all the cells. Cells then do one of two things with those little molecules: Put them together to make big stuff again through a process called anabolism or make energy though the process of cellular respiration (CR).

Catabolic reactions break things down, whereas anabolic reactions build things up. The cells take monosaccharides and build polysaccharides or carbohydrates; amino acids are bonded together to form polypeptide chains and proteins; and fatty acids are used to manufacture larger lipid molecules.

Cells can also use digested materials to make energy though the process of cellular respiration (CR). This is a four-step process that mainly takes place in the mitochondria, the cell's powerhouse. We'll look at this process in more detail later in chapter 6.

The energy produced is stored in the body's energy "molecule," ATP (adenosine triphosphate).

✔ **CONCEPT!** The body requires energy to do everything! Energy = ATP = Mitochondria. ATP = Energy = Mitochondria. Repeat out loud: Energy = Mitochondria = ATP! Remember REVV CCCAMM.

There it is, *physiology on a single sheet or POSS!* Our bodies are continually ingesting *big stuff*, breaking large molecules into small molecules, circulating the little stuff throughout the body to cells. The cells then either make big stuff from the little stuff or they make energy through cellular respiration. *SUPER Simple!* Because POSS is helpful in understanding cellular respiration and protein synthesis, we'll go through it again in chapter 6. REMEMBER, repetition is essential to learning!

NOTES

Review Questions

- How can the H_2O sheet be effectively utilized?
- What are the four biochemicals that make up the human?
- What are the four primary tissues in the human?
- What are the 12 organ systems that make up the human, and what are their functions?
- In what situations would you use the H_2O sheet?
- How will the H_2O sheet reduce memorization and increase understanding?

Activities

- Explain the H_2O sheet to a classmate or any other A&P student. See if you can get them to *appreciate* it.
- Create a metaphor for each of the 12 body systems. Make sure to describe what each system does both in the body and in your metaphor.

FIGURE CREDITS

Fig. 2.2: Copyright © by LaiaMartínezM (CC BY-SA 4.0) at https://commons.wikimedia.org/wiki/File:Cos-huma-nivells-organitzacio-en.svg.

Fig. 2.3a: Copyright © by William Crochot (CC BY-SA 4.0) at https://commons.wikimedia.org/wiki/File:Les_Pagodes_de_Beauval_-_474.jpg.

Fig. 2.3b: Copyright © 2011 Depositphotos/Paha_L.

CHAPTER THREE

Body Systems and the Requirements for Life

TOES (TOPICS OF EMPHASIS)

- Definitions of anatomy and physiology and how they relate
- The integration of body systems
- Homeostasis—Dynamic constancy (DC) of the body (the big "H" word)
- Requirements for life and survival needs

Definitions

Anatomy is the study of human structure or form. Physiology is the study of human function. It is essential for any A&P student to learn these two definitions as we start to build a foundation of understanding.

In addition, A&P are complementary and arguably inseparable. As discussed earlier, form follows function. Structures in the human body are shaped the way they are because of what they do. This is the basic recurring concept in *SS A&P*. Strive to understand it to deepen your insight into anatomy and physiology.

✔ **CONCEPT FFF** Anatomical structures are shaped the way they are because of what they DO!

NOTES

Body Systems

The fourth column in the H_2O sheet lists 12 organ systems and abbreviates their conceptual function. Since *SS A&P* approaches A&P from a systems perspective

BRIDGE to the H_2O sheet to review the body systems

BRIDGE to Chapter 2, pages 15–17 to review body systems

it will be helpful to elaborate on those systems again. Some A&P reference texts will list a different number of systems, but we'll stick with 12 systems, which is a standard number.

Integumentary system—The skin is the only body system that comprises only one organ. It is also the largest organ *of* the body, not *in* the body. (The largest organ *in* the body is the liver in the abdominal cavity.) The skin *contains* our body fluids (mostly water) and provides *protection* as part of the innate immune system.

Muscular system—This system consists of our skeletal muscles, made up of striated or voluntary muscle tissue. Working with the skeleton, it allows for body movement.

Skeletal system—The bones provide support and protection and supply the framework for skeletal muscle attachments across joints. The skeletal and muscular systems work collaboratively and are sometimes considered a single system: the musculoskeletal system.

Nervous system—This is the body's primary communication system. It includes the central nervous system (brain and spinal cord), the peripheral nervous system, including sensory receptors and spinal nerves, and special senses like the eye and ear.

Endocrine system—The endocrine system is the body's chemical communication system, and helps to control and coordinate body functions. It functions by producing a variety of endocrine hormones like thyroid hormone, growth hormone, and the steroid hormones, estrogen, progesterone, and testosterone.

Cardiovascular system—The cardiovascular system or CVS is the body's transportation system. With some exceptions (general concepts usually have an exception), when things move from one place to another in the body it involves transport in the blood stream. And just like any transportation system the CVS includes a vehicle (blood), an engine providing energy (heart), and highways (vessels) where the vehicles move.

Lymphatic system—This system has a primary immune function but also acts as a tributary to the CVS by collecting interstitial (between cells) fluids and draining them back into the CVS.

Urinary system—The urinary system provides an interface between the CVS and the urine production tubules. It is the primary system responsible for maintaining fluid balance in the body while filtering wastes from the blood stream. Because it involves transport and fluid balance, it can be considered a system that collaborates with the CVS.

✔ **CONCEPT** Digestive System is a good example of the rule of two

Digestive system—The digestive system consists of two parts, the alimentary canal or gastrointestinal (GI) tract and six accessory organs, like the liver and pancreas. The digestive system absorbs nutrients from ingested foods and breaks those nutrients down so they can be absorbed and circulated throughout the body by the CVS.

Respiratory system—Through the process of gas exchange, this system provides oxygen essential for cellular energy production. It also removes (expires) carbon dioxide, one of the two waste products of cellular respiration.

Immune system—This is the body's defense system. It provides two types of immunity: innate and adaptive. We are born with the structures that provide innate or general immunity, like the skin and mucous membranes. Adaptive immunity allows our bodies to recognize "non-self" invaders (e.g., bacteria, viruses and fungi) and respond with the production of antibodies, which attack and destroy the foreign substances.

Reproductive system—The reproductive system allows the human species to continue by producing offspring. However, it is the one system that is not necessary to sustain the life of the individual. But it has great significance in the clinical setting.

NOTES

Homeostasis

This is the most essential concept of physiology, the body's magical ability of maintaining a relatively constant internal environment despite changes in the external environment. It can concisely be expressed as **dynamic constancy** or just **DC**.

✔ **CONCEPT** Homeostasis = Dynamic constancy or DC.

The human maintains homeostasis primarily through what are called negative feedback loops. These loops are just like an HVAC (heating, ventilation, air conditioning) unit in a home or building. HVAC systems contain a sensor (thermometer), an integrating center (thermostat), and an effector (something to heat or cool). A negative feedback loop *reverses* the original stimulus or signal. So, with an HVAC system, when the temperature rises a degree or two above the set point (maybe 70 degrees F) the system turns on the AC and cools the room down. If the temperature falls below the set point, then the system heats the room back up. The human body conceptually functions in a similar fashion with sensors, integrating centers, and effectors.

Human homeostasis will be discussed in detail as a basic concept of physiology throughout *SS A&P*, especially in chapter 13, The Endocrine System. An effective way to visualize (REVV CCCAMM!) homeostasis or DC is with a sine wave, as shown in Figure 3.1.

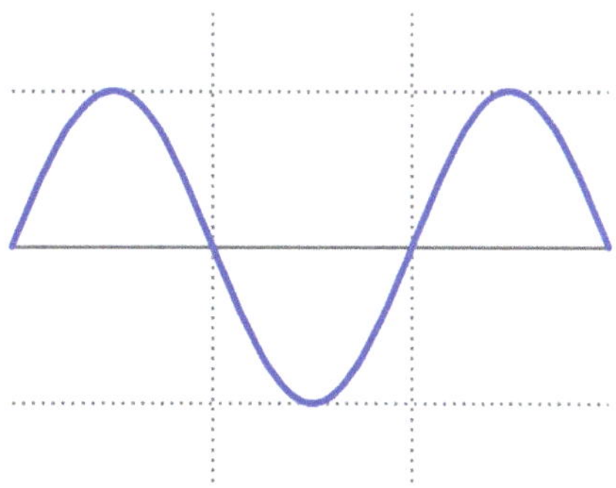

FIG. 3.1 Sine wave showing variation within upper and lower limits—DC

Requirements for Life

The introductory A&P student needs to consider two final concepts before moving on to a basic overview of the body. First, the requirements for life and second what the human body needs to survive. Humans are multicellular organisms where all body cells are interdependent on each other and the needs of the organism are filled by the body systems working cooperatively. Living organisms, including the human, require eight necessary functions, which are fulfilled by the structure of the organism. Remember, things are structured the way they are because of what they do. The following required functions for life shape how the human is structured. FFF over and over again!

Boundaries—Organism must maintain boundaries to separate the internal environment from the surrounding external environment. Human cells have a plasma or cell membrane, which creates a distinct boundary. Our skin provides a boundary, separating internal from external.

Responsiveness—Organisms need the ability to sense changes in the external environment so they can respond appropriately. Our nervous system is excitable and allows us to react to both external and internal stimuli.

Movement—The human must have the ability to move in order to live. Our skeletal muscle system, together with the skeleton, provides an ability to move. In addition, the contraction of visceral (smooth) and cardiac muscle moves substances within the body such as blood and digested food, nutrients, and wastes.

Digestion—In order to respond and move, our cells must produce energy. This requires an ability to ingest food, break those foods down, absorb the nutrients, and circulate them throughout the body via the CVS.

Metabolism—Organisms must have the ability to break down (catabolism) substances into smaller molecules. For example, the breakdown of carbohydrates (polysaccharides) into monosaccharides like glucose. Glucose is then used by the cells to produce energy (stored as ATP) through the process of cellular respiration. Or cells can utilize smaller molecules, like amino acids, to make larger macromolecules like proteins through the process of protein synthesis. These are anabolic reactions. Together catabolic and anabolic reactions make up cellular metabolism.

Excretion—Processes such as digestion and cellular metabolism produce waste. Digestion produces feces, and cellular respiration produces water and carbon dioxide. Organisms must have the ability to remove these substances through the process of excretion.

Reproduction—This process of duplication occurs at both a cellular and an organismal level. At the cellular level cells reproduce through mitosis. At the level of the organism, specialized cells in the ovaries and testes produce gametes (eggs and sperm). Those gametes can join and produce a single cell zygote, which will grow into a human organism through the process of mitosis.

Growth—The process of increasing the number of cells through mitosis or increasing the size of individual cells (hypertrophy) allows the organism to increase in size and to repair damage.

NOTES

Survival Needs

All organisms are designed to survive, but life is fragile and requires five specific factors:

Water—This simple molecule (one oxygen atom and two hydrogen atoms) is the most important substance in the body, making up almost two-thirds of our body mass.

Nutrients—Nutrients come from the foods ingested. The three primary nutrients are carbohydrates, proteins, and lipids. These break down into smaller molecules, monosaccharides, amino acids, and fatty acids, which are used by the cells to maintain life.

Oxygen—The chemical reactions that produce energy in the cells are oxidative reactions, which require the presence of oxygen. Human cells can survive for only a few minutes without adequate amounts of oxygen.

Normal body temperature—The metabolic reactions in the body require a stable body temperature within a normal range. Temperatures outside the normal range, either too high or too low, are not compatible with life.

Atmospheric pressure—This is the pressure of the air around our body. It is essential for breathing (ventilation) and the normal exchange of gases in the lungs.

Review Questions

(Write answers on a separate sheet of paper or in the NOTE spaces.)

- What is anatomy?
- What is physiology?
- How do anatomy and physiology complement each other?
- Define homeostasis in two words.
- What are the eight requirements for life?
- What are the five survival needs?

Activities

(Utilize separate sheets of paper.)

- Draw a graph showing dynamic constancy. Color it to make it more attractive and impactful.
- Explain each of the requirements of life.
- List and explain the survival needs of the human being.

FIGURE CREDITS

Figure 3.1: Copyright © by Omegatron (CC BY-SA 3.0) at https://upload.wikimedia.org/wikipedia/commons/thumb/0/02/Simple_sine_wave.svg/1280px-Simple_sine_wave.svg.png.

UNIT 1

OVERVIEW OF THE HUMAN BODY

CHAPTER FOUR

A&P Basics

Nomenclature, Terminology, Planes, Body Cavities, and Regions

TOES (TOPICS OF EMPHASIS)

- Anatomical nomenclature and terminology
- Anatomical planes
- Anatomical position
- Abdominal quadrants/regions
- Open and closed body cavities
- Homeostasis

Anatomical Terminology and Nomenclature

A wide variety of resources are available for detailing A&P terminology and nomenclature. The *SS A&P* Active Learning component by Cognella will help students learn terminology. Going online will also reveal numerous helpful sites.

It will take time for the novice student to become familiar with these terms. When working with the electronic version of *SS A&P*, utilize the audio function to learn and confirm pronunciation. Many other online sites will also provide a pronunciation function.

To help guide you, here's a list of the *most essential* anatomical terms and their meaning.

Reminder: Voice medical terms when you are learning them. Remember REVV CCCAMM? Vocalizing is a key to effective learning!

Terms

- Anterior—toward the front of the body
- Posterior—toward the back of the body
- Dorsal—a synonym for posterior; back of the body
- Ventral—a synonym for anterior; front of the body
- Superior—upward or toward the head

- Inferior—downward or toward the feet
- Superficial—on the surface; the skin is superficial
- Deep—under the skin or inside closed body cavities
- Medial—toward the midline of the body
- Lateral—toward the outside of the body
- Proximal—closer to the trunk or the beginning of an extremity
- Distal—closer to the outside of an extremity

Abdominal Quadrants

See Figure 4.1 on four surface quadrants.

- Right upper quadrant—major organs = liver and gall bladder
- Left upper quadrant—major organs = stomach and spleen + left lobe of liver
- Right lower quadrant—major organs = appendix, cecum, ovary in female
- Left lower quadrant—major organs = sigmoid colon, ovary in female

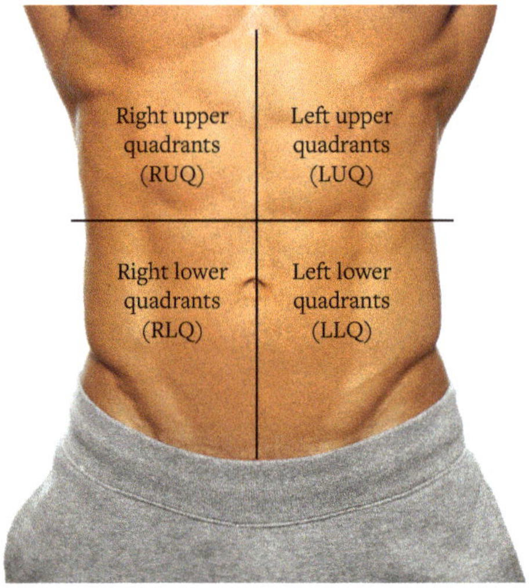

FIG. 4.1 Four abdominal quadrants

Abdominal Regions

See Figure 4.2 on abdominal surface regions; there are nine of them. What major abdominal organs lie under these surface regions?

- Right and left hypochondriac—under the ribs
- Right and left iliac or inguinal—in the area of the groin
- Right and left lumbar—at the level of the lumbar vertebrae
- Epigastric—above the umbilicus (belly button)
- Umbilical—region around the umbilicus
- Hypogastric—area below the belly button; pubic area

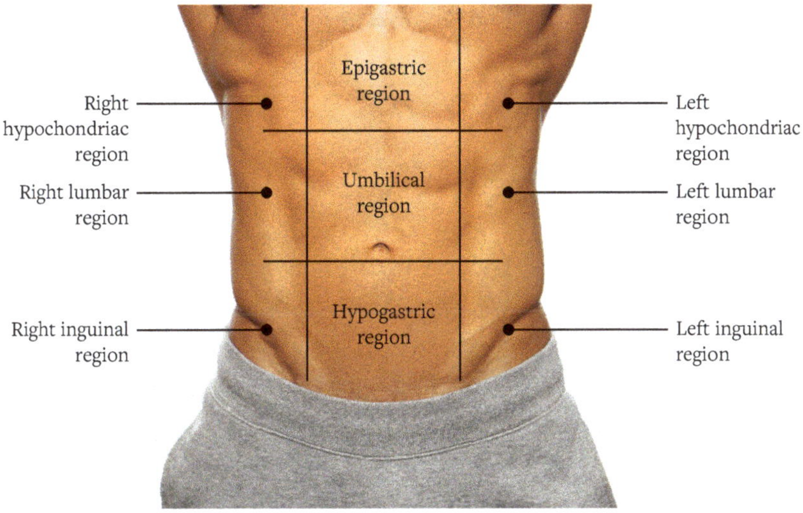

FIG. 4.2 Nine abdominal surface regions

The midsagittal plane is sometimes referred to as the median plane. Utilize a traditional A&P reference text, search the internet, or discuss anatomical planes with a study buddy or an instructor to make sure you understand.

NOTES

Go to Cognella Active Learning or other online sites and polish your understanding of anatomical terminology. As with any endeavor, "practice makes perfect"—or at least creates greater confidence! REVV CCCAMM!

Open and Closed Body Cavities

An open body cavity is any body cavity open to the outside environment. The oral cavity, nasal cavity, and rectum are examples. Open body cavities are lined by muscous membranes that produce mucus, which are a part of the body's immune system.

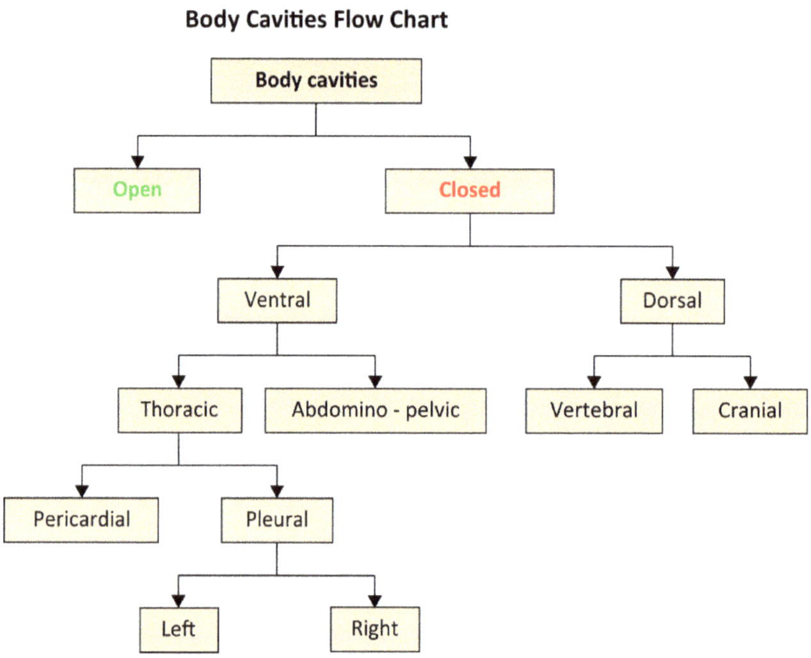

FIG. 4.3 Flow chart showing body cavity divisions

A closed body cavity is any body cavity closed to the outside. There are two major closed body cavities, the dorsal and the ventral cavities. These cavities are lined by serous membranes that produce a thin, watery, lubricating fluid, which facilitates the movement of internal organs like the heart, lungs, and intestines.

The dorsal cavity is divided into two subdivisions, the cranial cavity (A in Figure 4.4) that contains the brain and the vertebral cavity (B in Figure 4.4) that contains the spinal cord.

✔ **CONCEPT!** Notice the "rule of two" applies to closed body cavities. Frequently (not 100 percent of the time, but close) in A&P, one structure or item is divided into two subdivisions or categories. This concept will be seen as we continue to learn the fundamentals of A&P. Students can use this rule to minimize memorization!

The ventral cavity is divided by the diaphragm (main muscle of breathing) into the thoracic cavity and the abdominopelvic cavity (See Figure 4.4).

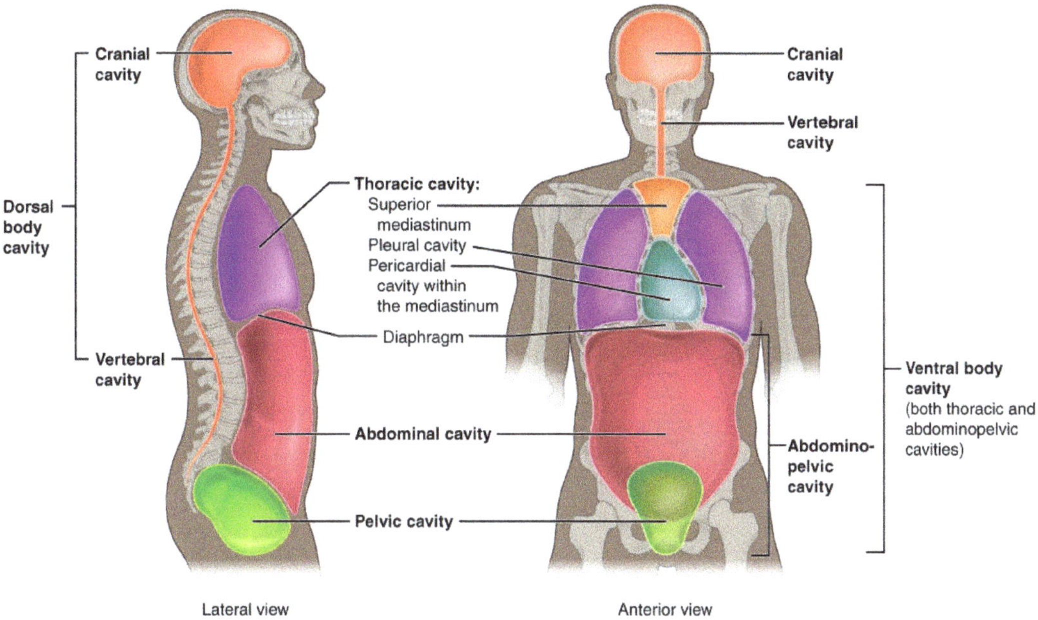

FIG. 4.4 Closed body cavities. Lateral and anterior views.

Serous Membranes and Mesentery

The serous membranes lining the closed body cavities cover both the organs and chamber walls of those cavities. Where the membrane touches the cavity wall, it is called **parietal**. Where the membrane touches the organ, it is called **visceral**. There are three ventral cavities where these membrane names apply:

- Pleura cavities, which contain the lungs
  - Visceral pleural—the serous membrane covering the lungs
  - Parietal pleural—the serous membrane lining the pleural cavity
- Pericardial cavity, which contains the heart
  - Visceral pericardium—the serous membrane covering the heart
  - Parietal pericardium—the serous membrane lining the pericardial cavity
- Abdominopelvic (AP) cavity, which contains all the abdominal and pelvic cavity organs
  - Visceral peritoneum—the serous membrane covering any of the AP organs
  - Parietal peritoneum—the serous membrane lining the AP cavity
- Mesentery – a double layer of parietal peritoneum suspending and supporting the viscera (organs or guts) of the G-I tract in the abdominal cavity.
  - mesentery proper – suspends most of the small intestine from the abdominal wall

- greater omentum – "fatty apron" hanging inferiorly from the stomach
- lesser omentum – runs between stomach, duodenum, and liver
- falciform ligament – attaches the liver to the abdominal wall
- mesocolon – attaches the large intestine to the abdominal wall

Help from Crash Course

An efficient and entertaining way to churn to learn A&P is "Crash Course" with Hank Green on YouTube! Just search "Crash Course A&P" on YouTube. Hank's A&P lessons will likely be one of the first sites that pop up. The lessons are an effective way to review and add insight into A&P. Each session is roughly ten minutes in length and moves quickly.

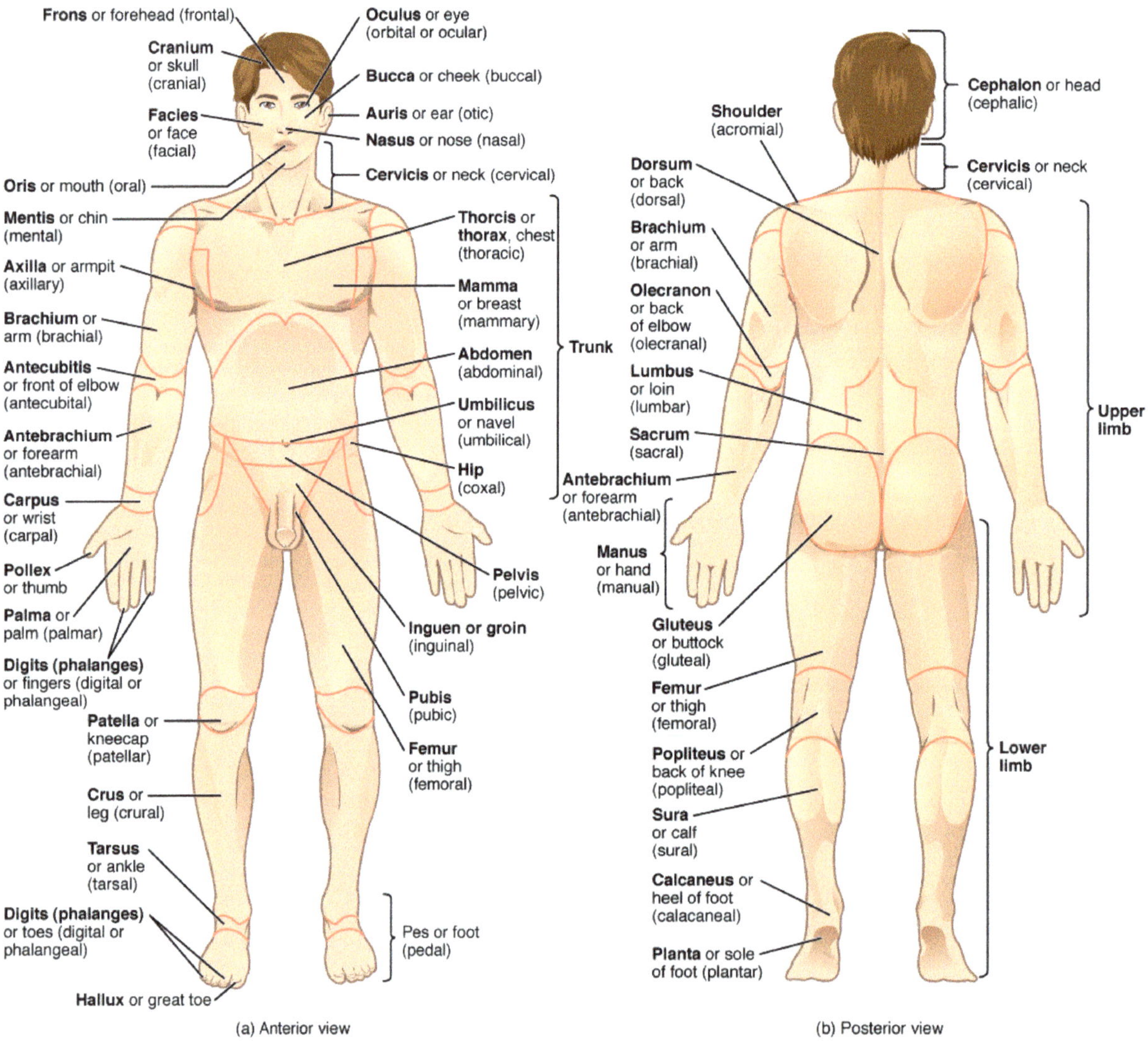

FIG. 4.5 Major anatomical regions of the body

Anatomical Regions

These are *essential* for students to learn. Body region terminology is utilized in clinical settings, and many arteries, veins, and nerves are named by the region of the body they occupy. Spend time churning Figure 4.5, which shows the most important regions the novice learner must grasp.

✔ **CONCEPT** Many arteries, veins, and nerves are named by the region they occupy.

Anatomical Planes

Anatomical planes divide or "slice" the body into different divisions or parts. There are three primary planes that the novice student must know:

- **Sagittal**—this plane divides the body into left and right halves and has two subdivisions
  - Midsagittal (also median)—divides the body into *equal* left and right halves
  - Parasagittal—divides the body into *unequal* left and right halves
- **Frontal** or coronal (yes, multiple names for the same thing!)—divides the body into front and back (anterior and posterior) parts
- **Transverse**—divides the body into upper and lower (superior and inferior) parts

Figure 4.6 demonstrates these planes.

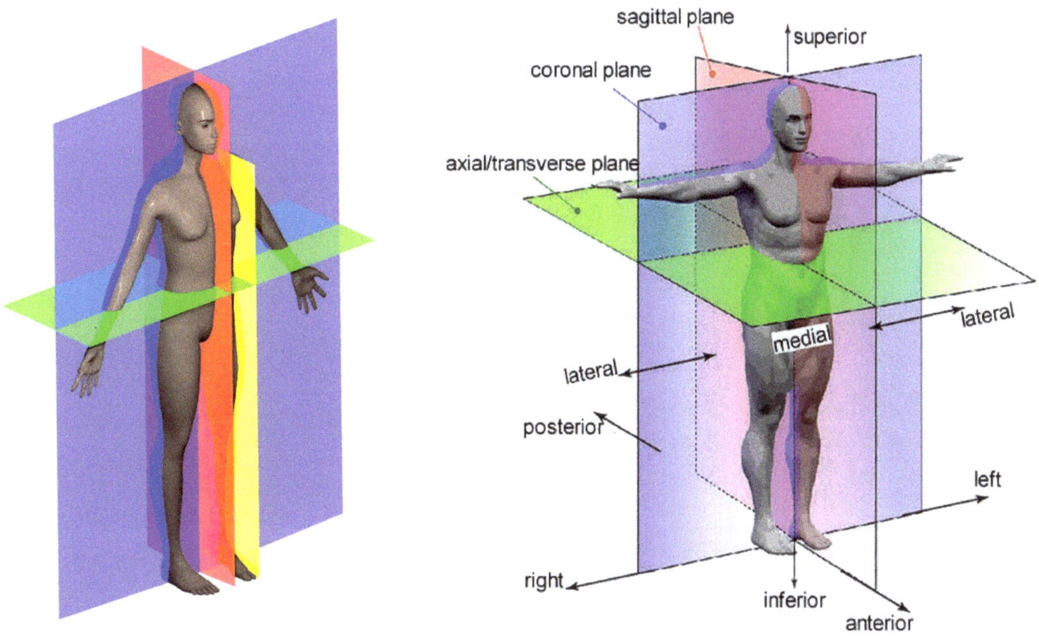

FIG. 4.6 Anatomical planes dividing body into portions

NOTES

Anatomical Position

It is important for any A&P student to know the term "anatomical position," what it means, and why it is important.

This is the standard reference position always used in A&P and medicine. We refer to the body as if it is standing erect, feet shoulder-width apart facing forward, with the upper extremities hanging at the sides. The palms are facing forward (anterior) so the thumbs are pointing toward the outside of the body or laterally.

Using the anatomical position eliminates confusion about the orientation of human anatomy. The head is always superior to the feet even if the body is inverted in a position where the feet are elevated above the head. We still refer to the head as if the body was in the anatomical position or superior to the feet. The thumb is always lateral to the little finger and the elbow inferior to the shoulder.

✔ **CONCEPT** We will *always* refer to the body as if it is in the anatomical position.

NOTES

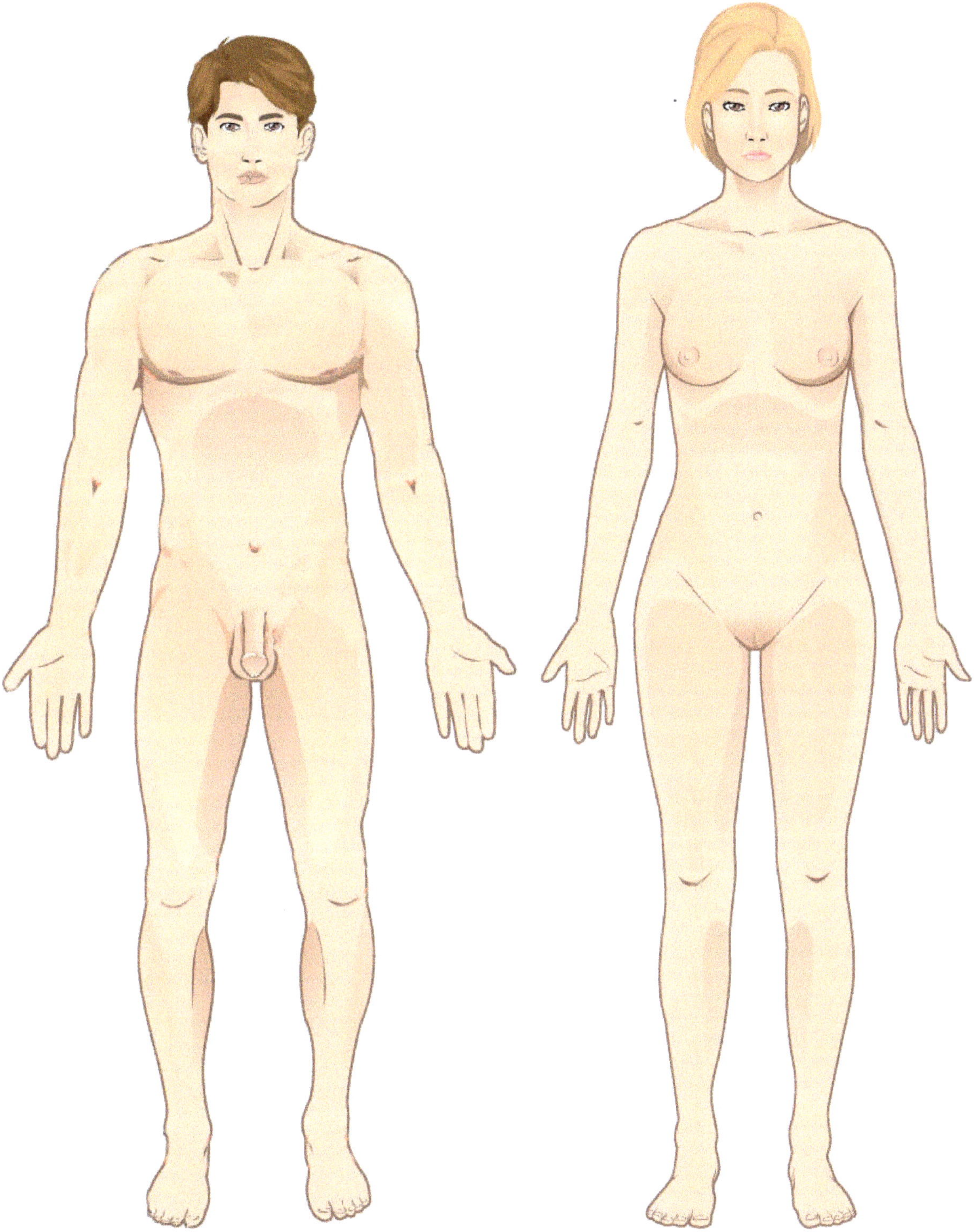

FIG. 4.7 Anatomical Position

Review Questions

- What do the terms “anterior” and “posterior” mean?
- How do closed body cavities differ from open body cavities?
- What’s the difference between a midsagittal and a median plane?
- What are the three anatomical planes and their subdivisions?
- What is the anatomical position, and why is it important?

Activities

- Make a list of 12 different ways to express FFF. Do this by changing the nouns (anatomy and physiology) and the verb. For example, form follows function can be expressed as anatomy follows physiology. Or, physiology determines anatomy.
- Add three more items to the FFF examples list you started at the end of chapter 1.
- List at least three organs that lie under each of the four abdominal quadrants (RUQ, LUQ, LLQ, and RLQ) Google this or utilize a reference text if you are having trouble.
- Draw a picture (to the best of your ability) of the body in the anatomical position. Have some fun with this one.

FIGURE CREDITS

CHAPTER FIVE

Chemistry Basics for Human Physiology

TOES (TOPICS OF EMPHASIS)

- Atoms, molecules, and ions
- Bonding—covalent, ionic, and hydrogen
- Molecular structure of water (H_2O)
- Acids, bases and pH
- Water mixtures
- Organic macromolecules (C, P, L, N)
- Enzymes

Chemistry Necessary to Understand Physiology

It is essential to grasp the basic concepts of chemistry to understand human physiology. But A&P is *not* a chemistry course, so we'll not go into any significant detail about chemistry. That journey is the responsibility of our colleagues in the Chemistry Department.

Chemistry is the science dealing with the composition and properties of substances and how they interact with each other. Go back and look at the H_2O sheet.

BRIDGE to Figure 2.1 to view the full-page H_2O sheet.

Look at column one, "Organizational Hierarchy." Remember, the human organism is developed by joining little stuff with little stuff to make bigger and bigger and bigger stuff, until there's a complete organism. Chemistry involves the smallest of the little stuff. We start with SAPs, or **S**ub**A**tomic **P**articles: protons, neutrons, and electrons. These three particles make up all the elements or atoms in the body.

An atom is the smallest particle, which exhibits the chemical properties of an element. Hydrogen is the lightest and smallest element (atom), whereas uranium is the heaviest and largest naturally-occurring element. The chemical elements (atoms) making up the human body fall into three categories: major, minor, and trace elements. These categories are based on the percentage of weight each element contributes to the body. The major elements make up more than 98 percent of the body's weight. The minor elements contribute less than one 1 percent of body weight. Trace elements make up less than 0.01 percent of body weight. There are only 12 elements that occur in the human body in more than trace amounts. So, those are the atoms we'll emphasize.

Atomic Structure

Atoms are structured with a central nucleus comprising protons and neutrons. Protons have a positive charge. That's easy to remember because **p**roton starts with a "**p**" and so does **p**ositive. Neutrons don't have any charge. They are *neu*tral thus the name "*neu*tron." Once again, the name tells you something!

✔ **CONCEPT!** A&P are logical sciences and many times the names of structures or processes reveal something about form and function. It's the concept of commonsense. Use this concept to gain understanding and minimize memorization!

Electrons have a negative charge and can be visualized as rotating around the nucleus in orbits (like the planets of the solar system) or as a "cloud" of energy. The electrons which occupy the outer-most orbits are those electrons that give the element/atom their reactivity, that is, the ability to bond or join with other elements to form molecules. These are called *valence electrons* and occupy the valence shell. We'll discuss the three types of chemical bonds shortly.

Major and Minor Chemical Elements Essential in Physiology

The six ***major*** elements are as follows:

- **Carbon**
- **Oxygen**
- **Hydrogen**
- **Nitrogen**
- **Calcium**
- **Phosphorous**

The six *minor* elements are as follows:

- Sodium
- Chlorine
- Potassium
- Sulfur
- Magnesium
- Iron

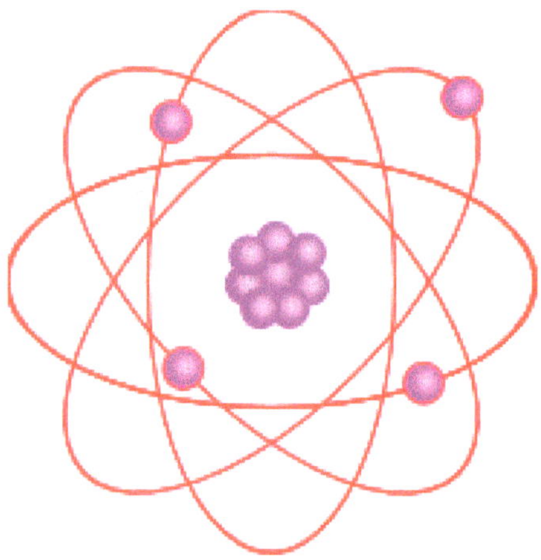

FIG. 5.1 Model showing a typical chemical atom

Chemical Elements Are Displayed in the Periodic Table

The periodic table is a tabular chart of all chemical elements arranged by atomic number, electron configuration, and recurring chemical properties.

The atomic number is determined by the number of protons in the nucleus of an element. The atomic weight is the sum of the number of protons and neutrons in the nucleus. Because the weight of

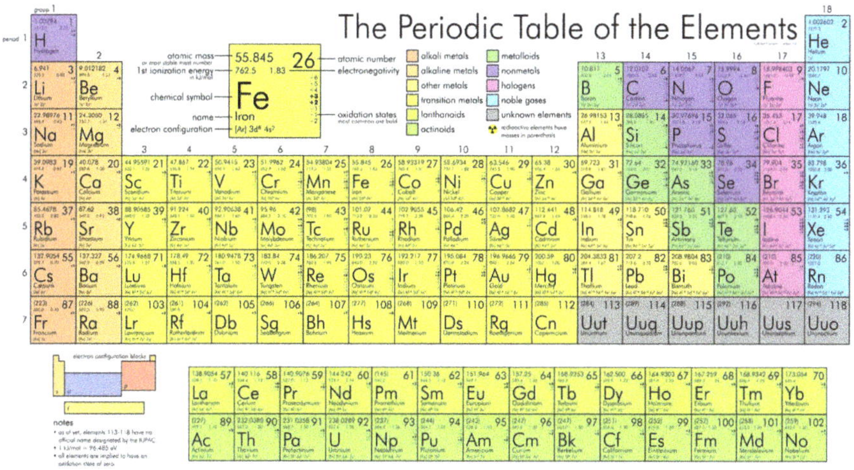

FIG. 5.2 Chemical periodic table

the electrons is so small, we will ignore them for this course. We'll only be concerned with the sum of protons and neutrons to create atomic weight.

Hydrogen is the number one atom, with an atomic number of one and an atomic weight of one because it has one proton, no neutrons and one electron. Hydrogen is a very important element in human physiology, as is the hydrogen anion, or free proton, which determines acidity in the body.

Carbon is one of the most essential atoms in human physiology. Substances containing carbon are referred to as **organic molecules**, whereas those molecules without carbon are called **inorganic**. Most of the organic molecules are carbon-based containing a carbon ring, either a five-sided pentagon ring or a six-sided hexagon ring, referred to as a benzene ring. Although there is some controversy among chemists, we will consider any molecule containing carbon, with or without a carbon-based ring, as an organic molecule. This includes two carbon oxides: carbon dioxide (CO_2) and carbon monoxide (CO).

BRIDGE to page 47 for details about pH.

✔ **CONCEPT!** Molecules that contain carbon are organic. Molecules without carbon are inorganic!

Ions are another chemical form essential in human physiology. An ion is an electrically charged particle. This includes electrically charged atoms, as well as molecules. Ions that carry a positive charge are called cations, and those with a negative charge are called anions. This can easily be remembered by visualizing the following:

Cation with a lower case "t" where the top of the "t" looks like a positive (+) sign.

Anion where the cross bar in the capital letter "A" can be visualized as a negative (–) sign.

OR

Cats have paws. Paws sounds like positive (another type of audionym!)

✔ **CONCEPT!** Ions are electrically charged particles!

Large protein molecules can carry an electrical charge, frequently a negative charge. Large protein anions play a significant role in maintaining the resting membrane potential (RMP) in cells, particularly neurons. The RMP is important in electrical impulse transmission in the nervous system. We'll discuss this further in chapter 12, when we cover the nervous system.

Isotopes

An isotope is an atom with the addition of one or more neutrons. Many but not all isotopes are radioactive. For example, adding a neutron to hydrogen produces a molecule of deuterium, adding two neutrons produces tritium. Tritium is radioactive, but deuterium is not.

"Heavy" water contains H_2O molecules where the hydrogen atoms are deuterium, containing a proton and a neutron in the nucleus. Remember, atomic weight is the sum of protons and neutrons, so deuterium really is heavier than hydrogen. Although heavy water is not radioactive, it is useful in the production of nuclear weapons. It was coveted by the Nazis during World War II when they were attempting to build an atomic bomb. Thankfully, Norwegian Resistance fighters gave their lives to prevent that from happening. WHEW! Just a little history to go along with our A&P!

Chemical Bonding

There are three types of chemical bonds. It helps to learn them in the order of bond strength, with the strongest first.

- ***Covalent bond***
  - Arguably the strongest chemical bond (some controversy about this)
  - The two bonding atoms "share" one or more electrons
  - Two types of covalent bonds (rule of two)

✔ **CONCEPT!** The rule of two keeps popping up over and over in A&P, just like two types of covalent bonds

    - Polar—electron(s) are shared unequally, causing electrical polarity
    - Nonpolar—electron(s) are shared equally: no electrical polarity
  - How atoms share electrons is the result of *electronegativity* (ability to attract electrons)
    - Atoms of the same element (O, H, C, etc.) have the same electronegativity
      - These atoms will share electrons equally
      - Equal sharing of electrons results in *nonpolar* molecules (O_2)
      - One oxygen and two hydrogen atoms join in a *polar* covalent bond to form H_2O or *water*, which is a polar molecule
    - Atoms of different elements have different degrees of electronegativity
      - These atoms will share electrons unequally

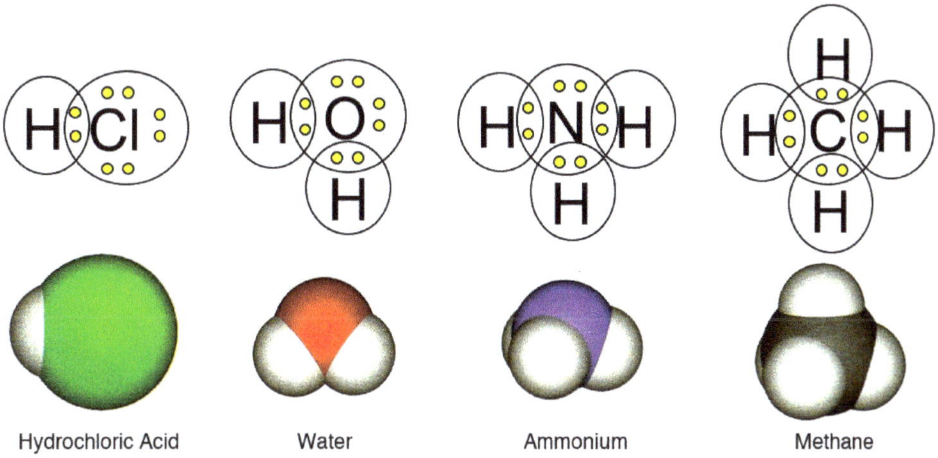

FIG. 5.3 Examples of covalent bonding sharing of electrons

- ***Ionic bond***—electron(s) are completely given up by one atom and absorbed by another atom. Common table salt (NaCl) is a good example of ionic bonding.

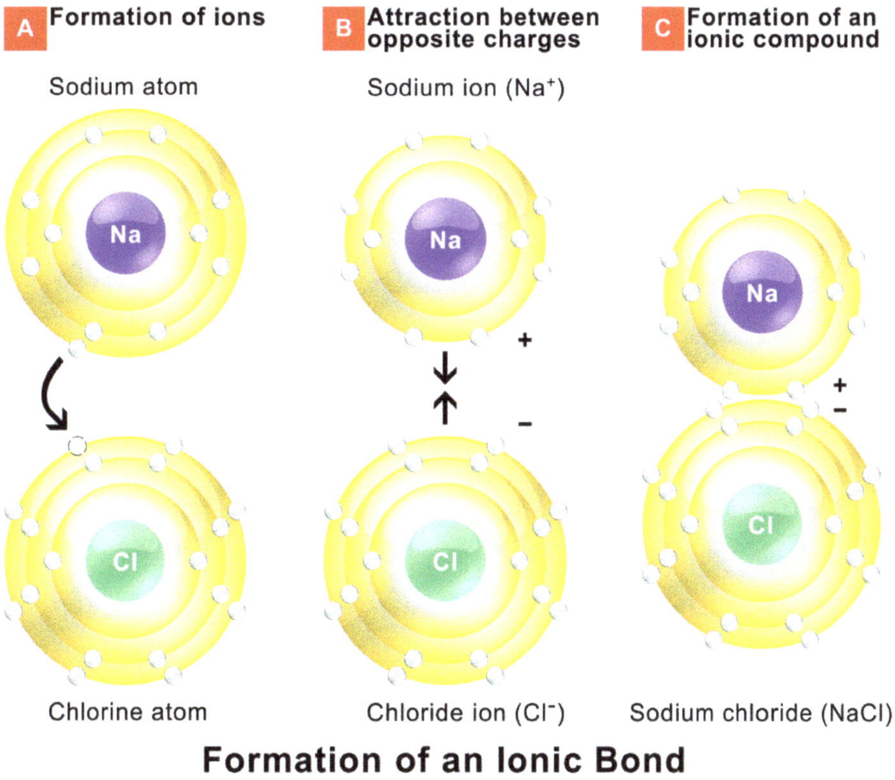

FIG. 5.4 Formation of an ionic bond in salt

- ***Hydrogen bond***
  - Weak attraction between polar molecules
  - Water (H_2O) molecules are polar (polar covalent bonds)
  - Hydrogen bonds between water molecules create *surface tension*

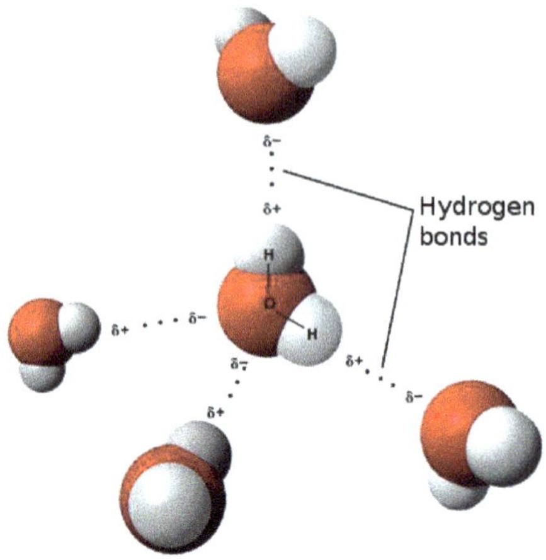

FIG. 5.5 Hydrogen bonding between water molecules

Water (H_2O)

Water is an essential molecule and has great importance in human physiology. It is a simple molecule comprised of only three atoms of two elements but accounts for roughly two-thirds (66 percent) of body mass. How solid molecules react in water is also important to understand when studying physiology.

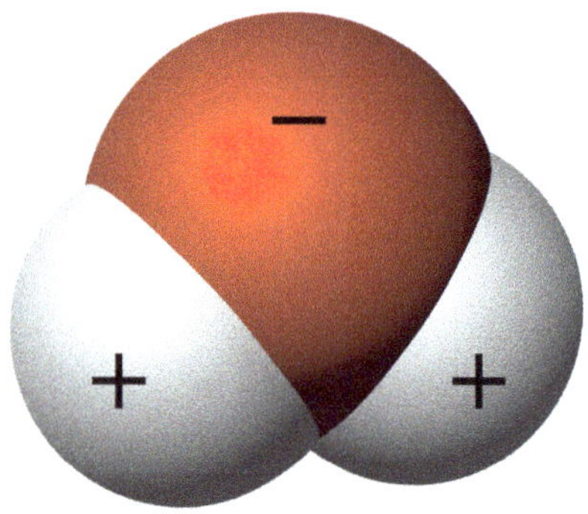

FIG. 5.6 Water molecule showing polarity

- Properties of water:
  - Water is present in three phases depending on temperature
    - Solid—ice
    - Liquid—water
    - Vapor or gas—steam
  - Almost all water in the body exists in the liquid form (Where would it be a vapor?)
  - Hydrogen bonds hold the water molecules together, preventing the water from becoming vapor
  - As a liquid in the body, water serves the following functions:
    - Transports—substances dissolved in water
      - Blood
      - Lymph
    - Lubricates—serous fluid in closed body cavities
    - Cushions—cerebral spinal fluid protecting the brain
    - Excretes—wastes such as urine
- **Surface tension**—attraction of polar water molecules through hydrogen bonding
  - Can be seen in the meniscus (curvature) of water in a test tube (Figure 5.7)
  - Allows water striders to glide across the surface of the water (See Figure 5.8)
  - Causes bubbles in pure water to collapse
    - Soap reduces surface tension, allowing bubbles to form in soapy water without collapsing.

- Surfactant protein produced by specialized cells in the alveoli in the lungs reduce surface tension (like the soap), which allows the alveolar sacs to stay inflated.
    - Lack of surfactant in premature infants (born at less than 26 weeks of gestation) causes "respiratory distress syndrome" because increased surface tension causes alveoli to collapse.

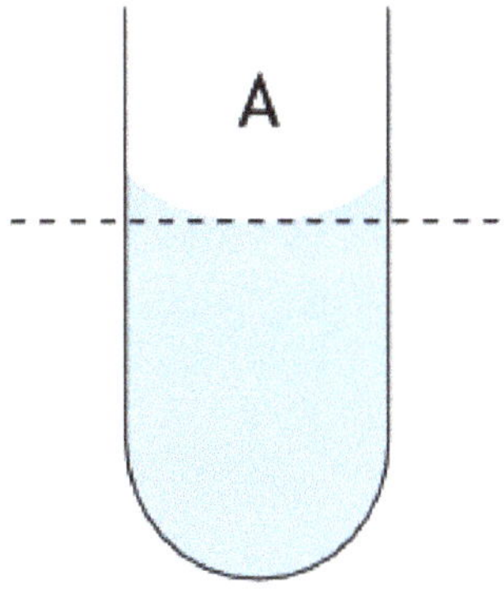

FIG. 5.7 Test tube meniscus caused by surface tension

- Water is the **solvent** of the body and molecules that dissolve in water are called **solutes**
    - Hydrophilic (like water) molecules dissolve in water—polar molecules and ions
    - Hydrophobic (dislike water) molecules do not dissolve in water—nonpolar
    - Water mixtures—formed by combining water and other substances
        - Three major types of water mixtures in human physiology:
            - Suspension—large particles that are held in the mixture by the motion of water
                - Sand in water
                - Blood cells in plasma
                - Particles can be separated from the mixture by filtration
            - Colloid—cloudy mixtures of large particles that remain mixed when not in motion
                - Gravy—large starch molecules in broth
                - Milk—lipid and protein particles in water
                - Cytosol—proteins in water
            - Solution—substances (solutes) disperse (dissolve) to make clear mixtures
                - Molecular solutes—polar molecules that can disperse in water (sugars, ethanol)
                    - Water surrounds solute molecules to pull them away from each other
                    - Solute molecules retain their identity in solution

FIG. 5.8 Water strider demonstrating surface tension from hydrogen bonding

- Ionic solutes—substances composed of positive ions (cations) and negative ions (anions) attracted to each other (salt–sodium chloride)
    - Water surrounds ions and pulls them away from each other (sodium ions—Na^+ and chloride ions—Cl^- interact with water molecules and move independently in solution)
    - Solutes can be recovered in their original form by evaporating the water

- Some materials will not dissolve in water and are **insoluble**
    - Molecular materials may not be polar enough to interact strongly with water molecules; nonpolar fats do not dissolve in water
    - Ionic materials may have attractions between ions too strong for water to pull them away from each other; calcium phosphate ($Ca_3(PO_4)_2$—hydroxyapatite) makes up the solid part of bone (the part that makes bones white) and is the substance we talk about when we say bones are made of calcium.

NOTES

Dehydration Synthesis and Hydrolysis Reactions

The building (synthesis) and breakdown of complex macromolecules are associated with two types of chemical reactions: dehydration and hydrolysis. Dehydration synthesis of complex biochemicals involves the joining of two smaller monomers (smaller particles) to make a dimer (larger molecule). During this process, a water molecule is produced and released.

Hydrolysis involves the breakdown of a dimer into two smaller monomers, with the addition of a water molecule. Figure 5.9 shows both dehydration and hydrolysis reactions in simple conceptual format.

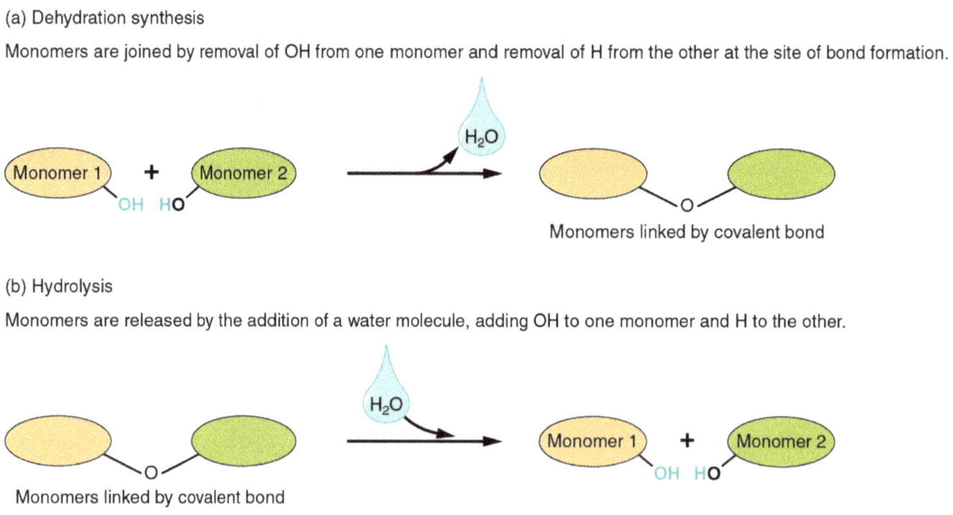

FIG. 5.9 Dehydration creates water; hydrolysis splits water

Acids and Bases—pH Scale

The balance of acidity in the human body is essential. The body has several mechanisms in place to maintain a narrow range of acid/base fluctuation. Acidity is measured utilizing the pH scale. Water readily dissociates into a hydrogen ion (H^+) and a hydroxide ion (OH^-). It is the concentration of hydrogen ions that causes acidity:

$$H_2O \rightarrow H^+ + OH^-$$

The pH of a solution measures the relative amount of hydrogen ions or free protons it contains. Hydrogen ions, or H^+, are the positive ions (cations) of the element hydrogen.

Since the hydrogen atom consists of one proton and one electron, to make H^+ the electron is removed, leaving only a proton. For this reason, H^+, hydrogen ion, and free proton are used interchangeably in the discussion of acidity.

pH is expressed as a number from zero to 14. Mathematically, pH is the logarithm of the inverse of the hydrogen ion concentration. Water has a neutral pH of 7. Any solution with a pH above 7 is a base and with a pH below 7 an acid.

BRIDGE to page 43 for details about ions.

✔ **CONCEPT!** pH measures the acidity of a solution (like plasma) caused by the concentration of hydrogen ions or free protons.

Slow down and take some time to get your head around this idea. It usually takes a while to understand. The pH logarithm is to the base ten, which means the difference in hydrogen ion centration between a pH of 7 and a pH of 6 is tenfold. It's a little like the Richter scale, which measures the intensity of earthquakes. To make it even more challenging to understand, pH is an *inverse* relationship. So, the more acidic a solution the *lower* the pH and the more alkaline (basic) a solution the *higher* the pH. One simple way to express the pH scale is as follows:

$$pH = \log \frac{1}{[H^+]}$$

This is one of those concepts you'll need to review more than a few times before you'll feel comfortable. However, the bottom line is a healthy body must maintain a homeostatic balance of pH ranging between 7.35. and 7.45.

pH	Examples of solutions
0	Battery acid, strong hydrofluoric acid
1	Hydrochloric acid secreted by stomach lining
2	Lemon juice, gastric acid, vinegar
3	Grapefruit juice, orange juice, soda
4	Tomato juice, acid rain
5	Soft drinking water, black coffee
6	Urine, saliva
7	"Pure" water
8	Sea water
9	Baking soda
10	Great Salt Lake, milk of magnesia
11	Ammonia solution
12	Soapy water
13	Bleach, oven cleaner
14	Liquid drain cleaner

FIG. 5.10 pH scale showing acidity of various solutions

Biological Macromolecules or Biochemicals

Arguably the most important chemical molecules are the organic biochemicals making up the body. Do you remember the difference between organic and inorganic molecules? This would be an excellent time to practice some repetition and review that difference. It has to do with carbon!

BRIDGE
to Figure 1.5 in chapter 1 to view the full-page H_2O sheet.

Column two in the H_2O sheet lists the four essential biochemicals of the human. These are carbohydrates (CHO), proteins (P), lipids (L), and nucleic acids (NA). The two nucleic acids are DNA and RNA.

When striving to remember these chemicals, always repeat them in the same order to facilitate the development of new neuronal synapses (bridge to page 171) to most effectively build learning "neuronets." Remember the importance of *verbalization* or *vocalization* in learning. So, vocalize these four molecules. Just say, "C, P, L, N ... C, P, L, N ... C, P, L, N" over and over to build new synapses. It's just like practice with a musical instrument or any athletic endeavor like hitting a golf ball. Practice will make perfect or at least better! But without practice there won't be much, if any, progress. *Remember,* REVV CCCAMM!

✔ **CONCEPT!** No practice = No progress!

The H_2O sheet lists the essential biochemicals in this order because it's the order of digestion in the human body. So, you're learning some important stuff about the digestive system when churning "C, P, L, N"!

These biochemicals contain carbon, hydrogen, and usually oxygen. Some will also contain nitrogen (proteins and nucleic acids), phosphorus, and sulfur. The carbon component may simply be an individual carbon atom or numerous carbon atoms arranged in what can be called a carbon "skeleton." The skeleton can be a chain, branch, or ring, like a benzene ring.

At a foundational level (remember, we're in the basement of our four-story building), it is not necessary to struggle to memorize the chemical structures of these four hydrocarbon molecules. Here's what's essential for you to know at an introductory level:

- **Carbohydrates**—abbreviated CHO because those are the only three elements they contain—carbon, hydrogen, and oxygen
  - Simple sugar monomers are called monosaccharides
  - Disaccharides are two monosaccharides bonded together
  - Polysaccharides have many monosaccharides bonded together
  - *Glucose* is an important monosaccharide
  - *Glycogen* (storage form of glucose) is an important polysaccharide
  - Sweet drinks like fruit juices and soft drinks contain sugars as monosaccharides and disaccharides
  - Foods like pasta, bread, and potatoes contain polysaccharides like starch
- **Proteins**—the "building blocks" of the body
  - Made from amino acids
  - Polypeptide chains
  - All cells produce some type of protein
  - Think of cells as tiny protein factories
  - Complex molecules that are key to human health

    - Fold into different shapes or structures
        - Primary
        - Secondary
        - Tertiary
        - Quaternary

- **Lipids—fats**
    - Diverse group of water-insoluble (hydrophobic) molecules
    - Store energy as adipose tissue
    - Significant component of cellular membranes (SPPLBL—*remember?*)
    - Make up steroid-based hormones like estrogen and testosterone
    - Triglycerides—most common lipid in the body
        - Long-term energy storage in adipose tissue
        - Cushions organs
        - Insulation—temperature regulation

✔ **CONCEPT!** Think of human cells as protein factories. They manufacture different proteins whose structure is derived from the genetic codes in the cell's DNA.

NOTES

Enzymes

Chemical reactions in the body must proceed at a rate fast enough to sustain life. The millions of chemical reactions in the body are facilitated by enzymes. Enzymes are biologically active catalysts that accelerate the speed of chemical reactions. That acceleration is essential for normal function of the body. Enzymes do *not* change anything about the reaction. They don't change the substrates (ingredients going into the reaction), they don't change the product (result of reaction), and the enzyme doesn't change. The only thing an enzyme does is speeds up the reaction.

Separate substrate particles bind with the enzyme molecule (a protein), which facilitates the chemical reaction to produce a product. The enzyme releases that product and is then available to catalyze another reaction. See Figure 5.11.

Most, if not all, chemical reactions in the body are under enzyme control. So, enzymes are essential for normal human function. Most enzymes are proteins and are manufactured through the process of protein synthesis by cells based on genetic codes contained in the cell's DNA.

✔ **CONCEPT!** Almost all chemical reactions in the body are controlled by **enzymes**. The only thing enzymes do is accelerate a reaction! That's it—they just speed it up!

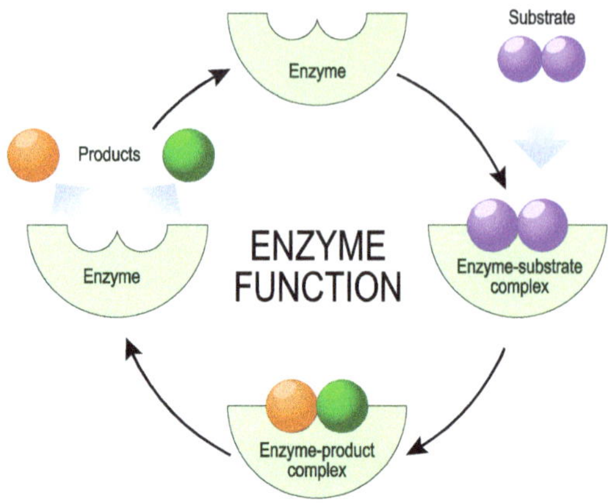

FIG. 5.11 Enzymes speed up reactions without changing substrates or enzyme

Oxidation/Reduction Reactions

Frequently the term "redox" is heard when studying chemistry. This refers to two reactions that are linked together, one is **oxidation** and the other is **reduction**. They are always paired together. In *redox* reactions, one molecule is "oxidized" by giving up electrons, while the other molecule is "reduced" by gaining electrons. Oxygen is frequently the recipient of the electrons, thus the term "oxidation. As we'll see in cellular respiration, glucose is oxidized by reducing two types of B vitamins, NAD and FAD, during the first stage of cellular respiration (CR).

✔ **CONCEPT!** When an atom/molecule is oxidized it gives up electrons. When an atom is reduced it gains electrons!

✔ **CONCEPT!** The mnemonic **OIL RIG** stands for "oxidation is loss, reduction is gain" and can help when remembering the difference between oxidation and reduction.

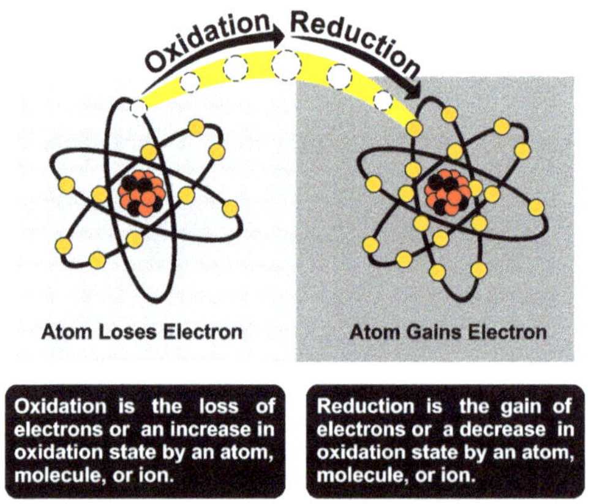

FIG. 5.12 Oxidation/Reduction Reaction—OIL RIG

Review Questions

- Why is it important to understand chemistry when studying physiology?
- What are the unique characteristics of water?
- What is a solvent?
- What is a solute?

- Why is a hydrogen ion also called a "free" proton?
- What is an enzyme, and why is it important in human physiology?
- Which chemical bond is the strongest? Which is the weakest?
- What is an ion?
- What makes water such a unique and essential molecule?
- Where in the body is water found in vapor form?
- What is the pH scale?
- What is an organic molecule?
- What is an inorganic molecule?

Activities

- Define and discuss the concept of pH. Write out the pH formula and explain the concept (logarithmic and inverse) to someone who is not familiar with the pH concept.
- List the three types of chemical bonds from strongest to weakest.
- Discuss the concept of C, P, L, N.

Coloring Activities

(Use a separate sheet of paper and/or space provided.)

- Using crayons, draw the pH scale with different colors representing levels of pH. Draw and color an example for each level of pH (e.g., water, milk, lemon, drain cleaner, etc.). Use Fig. 5.10 or just put "graphic of pH scale" into a browser to find examples online.
- Color a graphic showing the three different types of chemical bonds, including both polar and nonpolar covalent bonds. (Check online)
- Color a drawing demonstrating the function of an enzyme. (Use Fig. 5.11 or check online)

FIGURE CREDITS

CHAPTER SIX

Energy Production and Protein Synthesis

TOES (TOPICS OF EMPHASIS)

- Explanation of POSS
- Basic biochemicals
- Catabolism/anabolism
- Energy production in cells—CR
- Protein synthesis (PS)
- The process of PS—transcription and translation
- The importance of protein synthesis

Physiology on a Single Sheet (POSS)

The POSS was discussed in chapter 2, with a promise that we would review it later in *SS A&P*. Let's do that now.

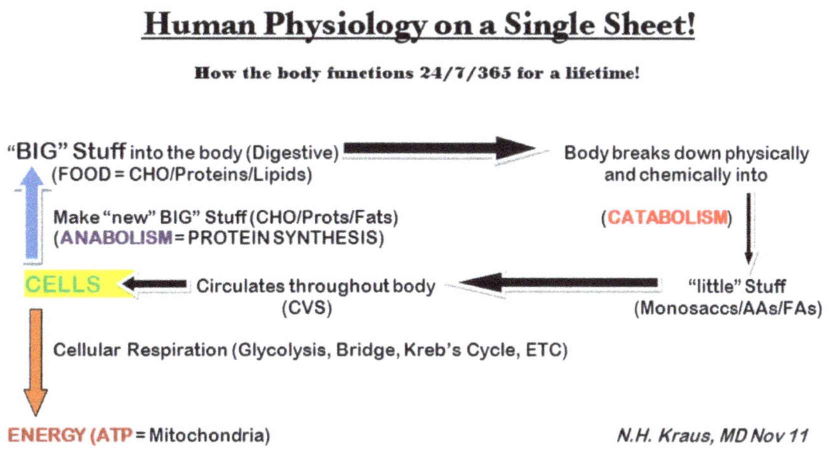

FIG. 6.1 Human function (physiology) on a single super simple sheet

Taking the time to understand this flow chart will save you hours of struggle later in any introductory A&P course, so let's dissect the POSS in detail and churn to learn it (understand it!) right now.

The POSS is a classic example of simplifying complex topics, another reminder that we're in the basement of the four-story building we're constructing. To build a solid foundation requires simplistic concepts. POSS is a simplified concept presented in the form of a graph, making it easier to visualize, remember, and utilize.

No matter what we're doing, our bodies are repeating the same activities needed to maintain human life. Look at the upper left-hand portion of the POSS. We must *eat*, ingest food and drink, on a regular basis to sustain life. We eat big stuff, like a taco. This contains the three basic sources of nutrition: carbs, proteins, and fats. Our food sources, both animal and plant, also contain nucleic acids (DNA and RNA), but these occur in relatively small amounts and so are not a major source of nutritional value. For the sake of simplicity, we'll ignore the nucleic acids as a source of nutrition and focus on carbs, proteins, and fats.

Those three nutrition sources are digested or broken down both physically (chewing and churning) and chemically (digestive enzymes). The process of breaking big stuff into little stuff is called **catabolism**. Chemical processes that break molecules into smaller pieces are catabolic reactions. The carbs are broken down to monosaccharides; proteins into amino acids; and lipids into fatty acids. Those small molecules are then circulated throughout the body by the cardiovascular system or CVS. Look at the fourth column on the H_2O sheet and find the cardiovascular system. Underneath the CVS is the word "transport." The CVS is the body's transportation system.

H_2O **BRIDGE** to Chapter 2, page 11 to review the H_2O Sheet

✔ **CONCEPT!** With an occasional exception, whenever our bodies need to move stuff (nutrients, wastes, hormones, etc.) from one place to another in the body it involves the CVS.

The digested little stuff from that taco (monosaccharides, amino acids, fatty acids) is circulated throughout the body to all the cells. The cell is the functional and structural unit of the human. Cells do one of two things with those little molecules broken down from our food.

Cells can take the little molecules and put them together to make big stuff again. This process is called **anabolism**. Catabolic reactions break things down, whereas anabolic reactions build things up. The cells take monosaccharides and build polysaccharides or carbohydrates; amino acids are bonded together to form polypeptide chains and proteins; and fatty acids are used to manufacture larger lipid molecules like steroid hormones (estrogen or testosterone). That's one thing cells can do with the little stuff.

The other thing cells do with the little stuff is make energy though the process of cellular respiration (CR). This is a four-step process we'll look at in more detail later in this chapter. Most of the energy produced by CR takes place in the mitochondria, the cell's energy plant. The energy produced is stored in the body's energy molecule, ATP (adenosine triphosphate), and is used to fuel the activities of the body.

✔ **CONCEPT!** The body requires energy to do everything! Energy = ATP = Mitochondria. ATP = Energy = Mitochondria.

Repeat out loud: Energy = Mitochondria = ATP! Remember REVV CCCAMM!

There it is, physiology on a single sheet, or POSS! Our bodies are continually ingesting big stuff, breaking large molecules into small molecules, circulating the little stuff throughout the body in the CVS to cells. The cells then either make big stuff from the little stuff or they make energy through cellular respiration. **SUPER simple!**

At a foundational level it really *is* just that simple. You'll be able to add greater detail and understanding as you move forward in your academic and professional careers. Right now, it's essential for you to build a solid conceptual understanding which the POSS allows you to do.

NOTES

Cellular Respiration or Metabolism (There's a Difference!)

Cellular metabolism (CM) is the sum of all chemical reactions in a cell, both catabolic and anabolic. Cellular respiration (CR) is just the catabolic part of metabolism involving the breakdown of nutrients and the process of producing energy in the form of ATP.

Our bodies require energy to do everything and anything! So, energy production is essential. It's important at a foundational level for A&P students to grasp the concept of cellular respiration. The POSS shows cellular respiration in a simplified form. Now we'll examine it in more detail, but only the amount of detail required at an introductory level.

✔ **CONCEPT!** Cellular metabolism involves all the chemical reactions in the cell, both catabolic (e.g., cellular respiration) and anabolic (e.g., protein synthesis)

Figure 6.2 shows a simplified diagram of cellular metabolism demonstrating how the body ingests food, inhales oxygen, and produces energy. It shows both catabolism and anabolism (production of biomass), so it ties nicely with POSS. Spend some time working to relate Figure 6.2 with POSS.

Cellular respiration begins with the ingestion of food and the catabolism of carbs, proteins, and fats into monosaccharides, amino acids, and fatty acids. The cells can then use any of these small molecules to make ATP.

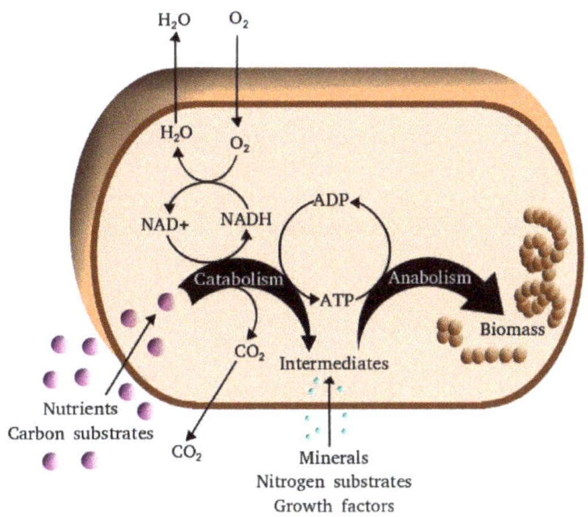

FIG. 6.2 Cellular metabolism involves both catabolism and anabolism

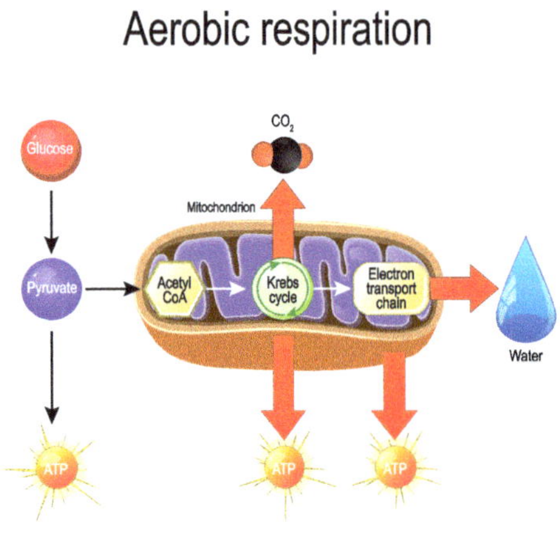

FIG. 6.3 Simplified graphic of aerobic CR

However, the primary source of cellular energy production is glucose, a simple sugar. Glucose is a monosaccharide derived from the breakdown of polysaccharides, such as starches, like bread and pasta.

Figure 6.3 shows a little more detail for CR. Although amino acids and fatty acids can also be utilized to create ATP, glucose is the primary source of cellular energy production. Glucose must go through the process of glycolysis, which takes place in the cytosol or cytoplasm.

Glycolysis breaks down glucose (six carbon atoms) into two molecules of pyruvic acid or pyruvate (three carbon atoms each). Glycolysis produces only a small amount of energy (2 ATP) but more importantly produces high-energy electrons, which are carried via a reduced form of B vitamin (NADH) to the final stage of CR, the electron transport chain (ETC).

The pyruvate passes into the mitochondria where it enters the Kreb's or citric acid cycle. This cycle produces additional ATP (four ATP molecules for every molecule of glucose) and high-energy electrons, which are carried by reduced forms of B vitamins NADH and $FADH_2$ to the ETC.

The final phase of CR is the ETC. In this stage, the energy from the high-energy electrons in the reduced NADH and $FADH_2$ is used to create ATP through the process of oxidation phosphorylation. This produces the majority of ATP in CR. CO_2 and H_2O are produced as waste products. The water will be excreted by the kidneys and the carbon dioxide exhaled by the lungs.

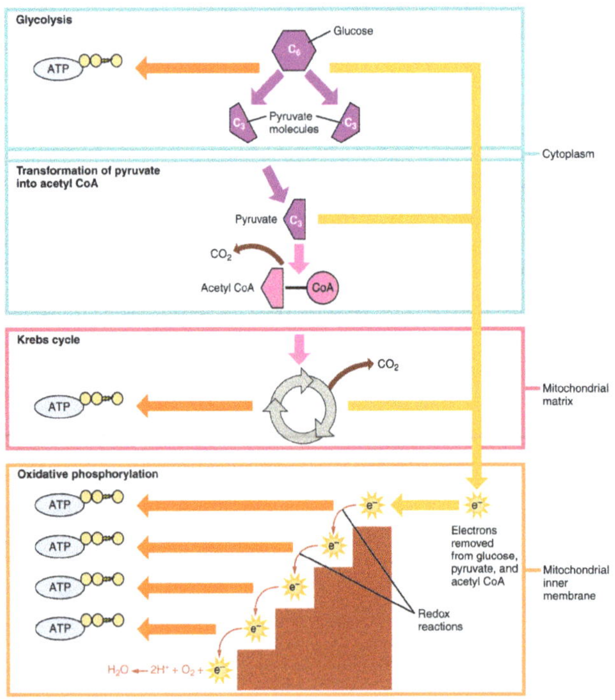

FIG. 6.4 Process of cellular respiration—glycolysis through ETC

✔ **CONCEPT!** Glucose **must** go through glycolysis to produce energy (ATP).

Glycolysis literally means the breakdown or lysis of sugar (glucose). The breakdown of glycogen (storage form of glucose) is glycogenolysis. Glycolysis takes place in the cytoplasm of the cells. It's a series of ten chemical reactions, but at this level, we're not concerned with those details. Just focus on the concept of CHOs breaking down into glucose, which goes through glycolysis, which is the first step in the four-step process of CR.

Figure 6.4 shows a more detailed yet still conceptualized diagram of CR, which allows us to review this energy production process a third time, but in more detail. The repetition will help deepen insight and improve retention. REVV CCCAMM!

NOTES

Glucose is a 6-carbon atom that breaks down into two 3-carbon atoms of pyruvate or pyruvic acid. Glucose starts as a 6-carbon atom and through glycolysis is split into two 3-carbon atoms of pyruvate.

Glycolysis requires a small amount (two molecules) of ATP to get the reaction started. This makes the first part of glycolysis an **endergonic** reaction. That is, a chemical reaction that *requires* energy to proceed. But as glycolysis moves through the individual chemical reactions it produces a total of four molecules of ATP, so, overall glycolysis is an **exergonic** reaction. That is, a chemical reaction that gives off energy in the form of ATP.

To summarize, glycolysis uses up two ATP to prime the pump but gives off four ATP. Four minus two results in a total production of two ATP from glycolysis. It's not as important for you to remember the specific numbers (2, 4, 2) of ATP in glycolysis. Rather, just grasp the foundational concept that glycolysis gives off a small about of energy.

What is important for you to retain is that glycolysis is the process of oxidation of glucose by the reduction of NAD (a type of B vitamin). This redox process of glycolysis produces a small amount of ATP, but, more importantly, the production of NADH, which contains or carries high-energy electrons. These will be vital to the production of large amounts of ATP in the final phase of CR.

Here's another place where it is beneficial to utilize a reference text or go online and see what you can find to help you gain insight into CR. **"Crash Course"** with Hank has an excellent segment on CR, and there are dozens of other tools that are helpful to the novice A&P student.

Finally, glycolysis is an **anaerobic** process. This means it does *not* require the presence of oxygen to function. But for the final three steps of CR to proceed, there must be adequate amounts of oxygen present in the cells. Otherwise, CR stops at the end of glycolysis. Without adequate oxygen, CR stops after the production of the two pyruvate molecules. To continue producing ATP, cells will convert the two pyruvate molecules into two molecules of lactic acid, by oxidizing NADH back into NAD. This is the process of **lactic acid fermentation.**

These two lactic acid molecules are then transported to the liver by the CVS. Remember the CVS is the body's transport system. When the body moves stuff, it usually puts the stuff (like lactic acid) into the blood stream for transportation.

The CVS carries the lactic acid to the liver, where it is converted back into pyruvate and then into glucose. Remember most, if not all, chemical reactions in the body are two-directional. These reactions will progress either to the left or right of a chemical formula depending on a variety of factors, including the concentration of substrates in the reaction.

So, the liver changes the lactic acid back into glucose and then releases it into the CVS for transport back to the skeletal muscles where the glucose can again go through glycolysis. This way the cells can continue to produce energy (ATP) through glycolysis. This is **anaerobic** respiration and is inefficient and cannot be sustained for very long. This is what a runner experiences when they say they've "hit a wall" in a marathon.

The process of converting lactic acid into glucose and circulating it from the liver back to skeletal muscles is called the **Cori cycle**. The production of glucose from pyruvate is called gluconeogenesis (making new sugar). The breakdown of glycogen (storage form of glucose) is

BRIDGE
to Chapter 7, page 72, to learn more about osmosis

glycogenolysis and the breakdown of glucose is called glycolysis. The liver (and skeletal muscles) store glucose in the form of glycogen. This is because glucose is highly osmotically active and glycogen is osmotically inert. Osmosis and osmotic pressure will be discussed in chapter 7.

Figure 6.5 demonstrates the Cori cycle in graphic form, helping you visualize the process.

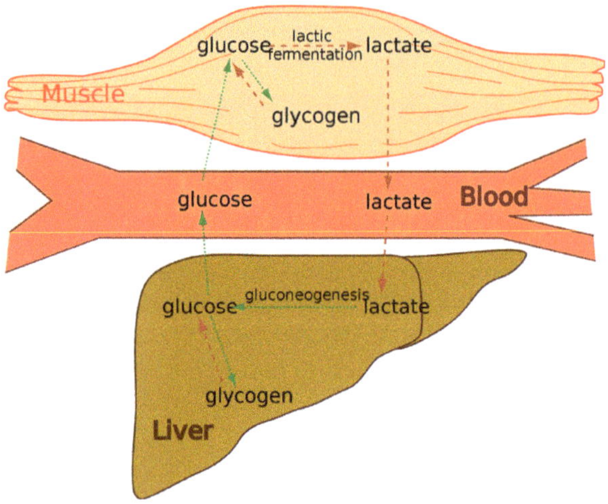

FIG. 6.5 Conversion of lactate into glucose in liver—Cori cycle

What does allow the cells to continue producing lots of ATP efficiently is aerobic respiration where there is adequate oxygen present. The presence of oxygen means the pyruvate produced by glycolysis can go through the bridge or intermediate stage (second step) of CR. This short process doesn't produce any ATP but does produce more high-energy electrons in the form of reduced NAD or NADH. Notice these high-energy electrons are carried as a part of a hydrogen atom in the reduced form of NAD. The intermediate step also produces CO_2 as a waste product.

NOTES

The end of the intermediate phase produces acetyl CoA, which is sometimes called the stargate molecule of CR. Do a search for the term "stargate" and see what you find. It was a movie and a TV show back in time, so the term may not be immediately familiar.

A stargate can be described as a wormhole, or an entry point to another universe or time and space. In CR, acetyl CoA is a common entry point into the third step of cellular respiration for all three of the nutritional molecules. Pyruvate from glucose, amino acids (AA), and fatty acids (FA) can enter the Kreb's cycle (third stage) through acetyl CoA.

✔ **CONCEPT!** Acetyl CoA is the stargate molecule of CR!

NOTES

Acetyl CoA enters the Kreb's or citric acid cycle, the third step in CR. The Kreb's cycle is a series of eight chemical reactions (again, the exact number is insignificant for our level of learning), which spins around twice for every glucose molecule processed. Remember, each glucose molecule (six carbons) produces two pyruvates that turn the Kreb's cycle *twice*!

The Kreb's cycle produces a small amount of ATP and a bunch of reduced NAD and reduced FAD. These reduced molecules transport high-energy electrons. NADH and $FADH_2$ are the most important products from the first three steps of CR, including the citric acid cycle. The Kreb's cycle also produces CO_2 as a waste product.

If there is adequate oxygen available, NADH and $FADH_2$ then proceed to the electron transport chain (ETC). The ETC takes place in the inner membrane of the double membrane surrounding the mitochondria. Remember the mitochondria? Repetition is appropriate here. The mitochondria are the cells' power plants. The first step of CR (glycolysis) takes place in the cytoplasm, but the remaining three steps all take place in the **mitochondria**.

✔ **CONCEPT!** Mitochondria are the power plants of human cells!

In the ETC, the energy from the high-energy electrons in the reduced form of NADH and $FADH_2$ is used to pump free protons or hydrogen ions from inside the matrix of the mitochondria into the space between the inner and outer membrane of the mitochondria. This creates a concentration gradient of free protons. Once the electrons have given up their energy to build this concentration gradient, oxygen accepts the electron and is reduced, bonding with hydrogen ion to form H_2O. Remember, the bridge and Kreb's cycle produced carbon dioxide (CO_2) as a waste product. So, the two waste products of CR are water and carbon dioxide.

✔ **CONCEPT!** CO_2 and H_2O are the two waste products of CR.

With adequate amounts of oxygen present to accept the spent electrons, the concentration of hydrogen ions (free protons) in the intramembranous space of the mitochondria flows through an integral enzyme called ATP synthase (-ase on the end of a chemical means it's an enzyme), located in the inner membrane of the mitochondria. The rush of free protons down its concentration gradient literally spins a rotor in the ATP synthase, which causes the bonding of a free inorganic phosphate group to ADP to form ATP. This process is much like water rushing through a tube in a hydroelectric dam (think Hoover Dam), which spins a turbine producing electricity. The ETC with ATP synthase is a similar process.

NOTES

Protein Synthesis (PS)

The production of energy through cellular respiration is essential for normal physiological function. Protein production is another crucial cellular process. CR is a catabolic reaction whereas PS is anabolic. Proteins are made from the twenty amino acids needed for normal cell function. Nine of these amino acids are essential, meaning they must be ingested in food. The other eleven can be produced in the body and are called non-essential.

Most human cells are little protein factories. Our cells make millions of different types of proteins essential for normal, homeostatic health. Proteins are complex molecules performing an almost limitless variety of physiological tasks. Cells make proteins through the process of protein synthesis. Synthesis means to make or manufacture, so here, again, the name tells you exactly what it is. But how do cells know the complex structure of these essential molecules? How are cells able to make these protein miracles?

✔ **CONCEPT** The concept of common sense tells us protein synthesis is the process of protein manufacture.

Traditional A&P textbooks discuss the details of protein synthesis involving the nitrogenous bases (adenine, cytosine, guanine, thymine, and uracil). But at a foundational level, it's not necessary to go into that type of detail. Remember, we're staying in the basement of our four-story building, pouring a solid foundation. The details of protein synthesis are first- and second-floor stuff. Let's stay in our basement and churn away at the conceptual foundation of PS.

NOTES

The nucleus contains **DNA**, a double-stranded nucleic acid containing **genetic codes**. These codes are like recipes carrying the instructions for the manufacture of big stuff (large molecules, remember POSS?), mostly proteins. The DNA has the recipe, but it's trapped because the DNA cannot get out of the nucleus. So, the cells copy (like a Xerox machine) the genetic code from the DNA into a molecule of messenger RNA or mRNA. This copying process is called **transcription**. mRNA is called messenger RNA because it carries the message of the genetic recipe.

✔ **CONCEPT** Messenger RNA or mRNA is another example of the concept of commonsense. The name tells you what mRNA does!

RNA is another nucleic acid like DNA but is single stranded. The mRNA carries the genetic code recipe from the nucleus out into the cytoplasm. In the cytoplasm, the mRNA joins with another type of RNA called ribosomal RNA or rRNA. These ribosomes are the intracellular organelles that string together the amino acid building blocks making proteins. The recipe from the mRNA instructs the rRNA, telling it the specific sequence of amino acids in the polypeptide chain. The amino acids necessary to make the proteins are transported to the ribosomes by a third type of RNA called transfer RNA or tRNA. This stringing together of amino acids is called **translation**. Translation produces the final polypeptide chain or protein. Technically, a polypeptide chain is not a protein unless it contains at least one hundred amino acids. But, to keep things super simple, we'll refer to all polypeptide chains as proteins. Insulin is a good example. It is made up of 51 amino acids in a polypeptide chain, which is twisted and held together by three disulfide bonds. But, for our purposes, it's a protein, a very essential protein.

Figure 6.6 shows protein synthesis in a simplistic form.

FIG. 6.6 Easy to understand graphic of protein synthesis

✔ **CONCEPT!** Protein synthesis (PS) is the process of making proteins. It involves two steps: transcription and translation.

NOTES

Review Questions

- What does POSS stand for?
- What is the difference between anabolism and catabolism?
- What two things do cells do with nutrients like glucose, AAs, and FAs?
- Glycolysis takes place in what part of the cell?
- The first three phases of CR primarily produce what particle that fuels the ETC?
- What are the two waste products of CR?
- What is the Cori cycle?
- Why is glucose stored in cells as glycogen?
- How are the waste products of CR removed from the body?
- Which stage of CR produces the most energy?
- What's the difference between aerobic and anaerobic respiration?
- How do cells manufacture proteins?
- What are essential amino acids?
- What is a genetic code?
- What is DNA?
- What is RNA?
- Why are DNA and RNA important in human physiology?
- What are ribosomes?

Activities and Coloring!

- Time to get out those colored pencils or your crayons again! Coloring is a *great* way to solidify your learning. Draw a colorful diagram of the four steps of CR. Do *not* put in any details, just the concepts, like Figure 6.2.
- Explain your CR concept diagram to someone who doesn't know A&P. Struggle to get them to grasp the concept of cellular energy production.
- Explain POSS, physiology on a single sheet, to a classmate. Struggle to teach the concept to another student or three! Remember, you'll always learn more attempting to teach things to others!

FIGURE CREDITS

CHAPTER SEVEN

Simplified Cell Biology

Organelles and Membrane Transport

TOES (TOPICS OF EMPHASIS)

- Definition of cell biology
- Structure of a typical generic human cell
- Structure and importance of the cell membrane
- Major intracellular organelles and their functions
- Structure of the nucleus
- Mitosis and meiosis
- Discussion of cellular membrane transport
- Definition of diffusion (simple and facilitated)
- Discussion of osmosis
- Solvents, solutes, and the lipid membrane
- Understanding tonicity

Cell Biology

The cell is the structural and functional unit of the human body. Cell biology is the study of the cell, intracellular structures, and their functions. It is essential to understand the basic structure and function of cells to comprehend human anatomy and physiology.

Typical Human Cellular Structure

It's essential for the novice A&P student to spend time learning the following basics about the human cell. Those items in **bold** are most important.

- Human cells are called eukaryotic cells—meaning our cells have a nucleus, cytoplasm containing intracellular organelles, and a cell membrane.
- Students are encouraged to repeat those three cellular parts in the same order (outside to inside the cell) each time to help build solid memory networks.
    - Cell membrane

- Cytoplasm
- Nucleus

✔ **CONCEPT!** The cell membrane is a semipermeable, phospholipid bilayer abbreviated as **SPPLBL**.

- **SP** = Semipermeable—some particles pass through; others don't
- **PL** = Phospholipid—molecules are fats with phosphate groups
- **BL** = Bi-layer—two layers, two phospholipid molecules thick

- **Cytoplasm or cytosol**—a viscous, syrupy fluid with lots of stuff suspended in it, things like the following:
  - Proteins
  - Carbohydrates
  - Lipids
  - Glucose
  - Amino acids
- Cytoplasm is like Jell-O but with raisins, sesame seeds, salt and pepper, and maybe even a little dirt from the kitchen floor mixed in. This analogy can help get a picture in your mind's eye of cytoplasm as a thick fluid packed with lots of stuff!
- **Nucleus**—the largest structure contained in the cytoplasm. It is generally considered the third major part of the cell, along with the SPPLBL and the cytoplasm.
  - Surrounded by a nuclear membrane
    - Similar in structure to the cell membrane or the SPPLBL
    - Also called the nuclear envelope
    - Protein molecules are embedded in the envelope forming pores
    - Pores have direct connection with the rough ER in the cytoplasm
  - Contains DNA—genetic material
    - Determines characteristics of the organism
    - DNA provides genetic "codes" for the manufacture of macromolecules, particularly proteins
  - **Nucleolus**—composed of protein and RNA; produces ribosomes
- **Intracellular organelles**—contained in the cytoplasm
  - Provide a variety of vital cell functions
  - **Cytoskeleton**—maintains the shape of the cell
    - Microtubules
    - Microfilaments
  - **Endoplasmic reticulum** (**ER**)—both smooth and rough

- **Rough ER** (lined with ribosomes)—produces proteins from genetic codes in DNA through the process of protein synthesis
- **Smooth ER**—makes lipids and detoxifies poisons

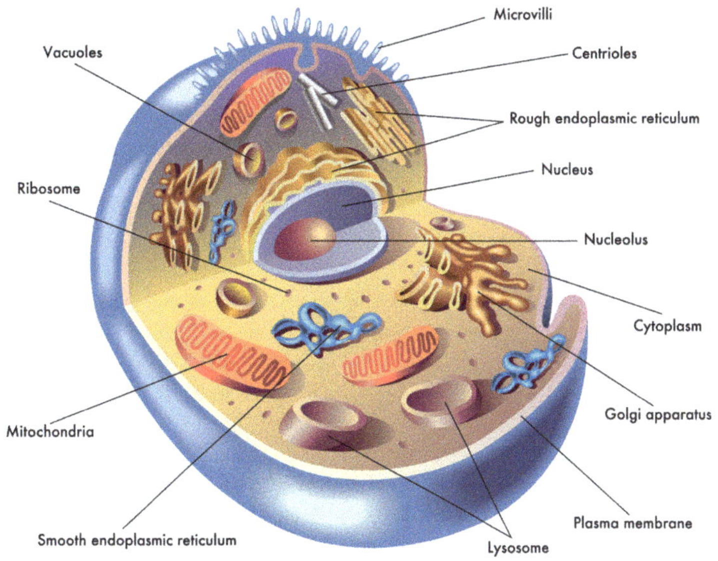

FIG. 7.1 Super simplified human eukaryotic cell

- **Golgi apparatus**—modifies and packages proteins produced by rough ER
- **Lysosomes**—break down ingested materials
- **Mitochondria** (the cell's energy plant)—CR produces energy stored as ATP
- **Ribosome** (on rough ER and free in the cytosol)—produce proteins
- **Centriole**—organizes microtubules during cell division (mitosis)
- **Vacuole**—fluid-filled sacs formed by the fusion of multiple vesicles

SPPLBL—Cell Membrane

The **SPPLBL** is a highly dynamic structure which can be described as being a "fluid mosaic." It is in constant motion and contains large integral protein molecules imbedded in the membrane. These proteins serve essential functions. For example, transport proteins provide channels for movement of particles across the membrane. Proteins can act as a physical barrier or serve as receptor sites for communication chemicals. About 20 percent of the SPPLBL is made up of **cholesterol,** which provides stability to the membrane.

> ***Cholesterol is an essential molecule in humans.*** It is necessary for life! It's received an unwarranted bad reputation. *SS A&P* will help restore its good name! More about cholesterol in later parts of this learning guide.

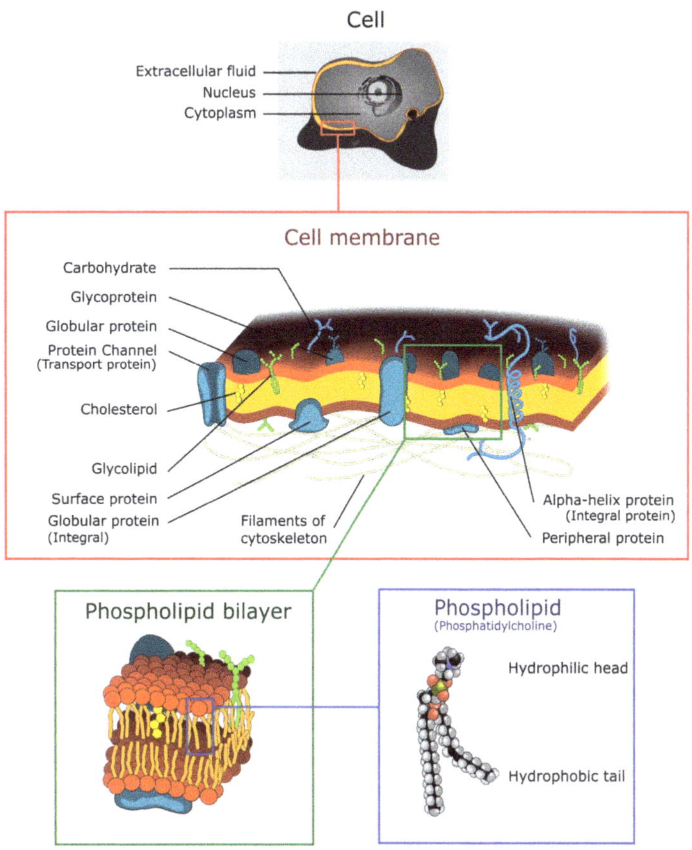

FIG. 7.2 Phospholipid bilayer cell membrane (SPPLBL)

To help visualize what is meant by the term *fluid mosaic* when describing the cell membrane, go online and put *"animation of fluid mosaic model"* in your browser. This will bring up a variety of animations demonstrating the cell membrane in motion. The human cell is fascinating!

The novice student can ignore the term *"alpha-helix"* in Figure 7.2 Just focus on the concept of the cell membrane containing integral proteins, which perform a variety of functions. Notice that the phospholipid (ignore the term "phosphatidylcholine," too much detail) molecule has a hydrophilic (likes water) head and a hydrophobic tail, which doesn't like water.

NOTES

Some human cells also have extensions of the plasma membrane, which allow the cells to perform specific functions.

✔ **CONCEPT!** Remember, things (including cells) are shaped the way they are because of what they do!

These cell membrane extensions include the following:

- **Microvilli** that increase surface area—facilitates absorption of nutrients in the small intestines
- **Cilia** are hair-like projects that wave and move to sweep material away from the cell, like the cells lining the bronchial tubes in the lungs.
- **Flagella**—a tail only found on sperm. Flagella provide motility for the sperm in their search for an egg to fertilize.

There are three basic ways human cells connect or hook on to each other. Once again, these cell junctions are structured the way they are because of what they do.

- **Tight junctions**—like spot welds holding cells tightly together. That's why they're called "tight" junctions. Tight junctions seal off the space between cells so substances must pass through the cell and not between cells.
- **Desmosomes**—provide added structural integrity needed when cells are under stress, like skin cells and cardiac muscle cells.
- **Gap junctions**—junctions with pores or holes in them; that's why they're called gap junctions. They allow substances (ions, glucose, amino acids, etc.) to flow directly from one cell to another cell. This is

FIG. 7.3 A&P so often make common sense

vital in a tissue like the heart where electrical activity is caused by the rapid movement of ions (electrically charged particles) resulting in a wave of depolarization leading to muscle contraction.

Finally, know the difference between mitosis and meiosis:

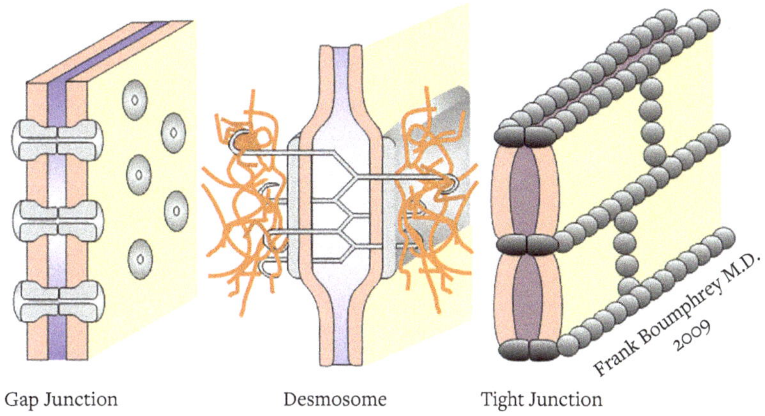

FIG. 7.4 Gap junctions, desmosomes, and tight junctions between cells

- **Mitosis** is cell division where one mother cell divides into two identical daughter cells
- Mother cell has a diploid number (46) of chromosomes
- Each daughter cell also has a diploid (46) number of chromosomes
- Mitosis can be divided into five stages
  1. Interphase—technically this is not one of the stages of mitosis. The cell spends most of its life in interphase. The first graphic in Figure 7.4 shows the very end of interphase as the cell begins to move into mitosis.
  2. Prophase—beginning of cell division; formation of chromosomes
  3. Prometaphase—nuclear envelope disintegrates; chromosomes migrate
  4. Metaphase—chromosomes line up across the center of the cell
  5. Anaphase—chromosome pairs break apart and move to opposite ends of the cell
  6. Telophase/cytokinesis—nuclear envelope re-established; cells separate

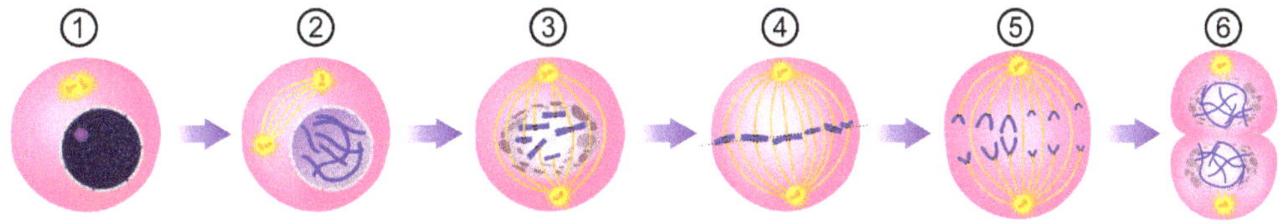

FIG. 7.5 Stages of mitosis providing growth and repair for the body

- **Mitosis** provides *two* functions for the body:
  - **Growth**
  - **Repair**
- **Meiosis** only occurs in the ovaries and the testes
  - One cell starts with a diploid number (46) number of chromosomes
  - Produces four sperm, each with a haploid number (23) of chromosomes
  - Produces one oocyte (egg) with a haploid number (23) of chromosomes
  - Allows for the formation of one diploid cell (zygote) when a sperm fertilizes an egg

NOTES

Understanding Cellular Membrane Transport

Cells are the basic structural and functional unit of the body. For cells to function, substances must be able to move from extracellular space (outside the cell) to intracellular space (inside the cell) and vice versa. There are a variety of mechanisms that accomplish this movement through the cell membrane (SPPLBL). Figure 7.6 summarizes what is important for the novice A&P student to know. First, we'll examine the passive processes not requiring energy. These passive processes move particles/molecules down a concentration gradient.

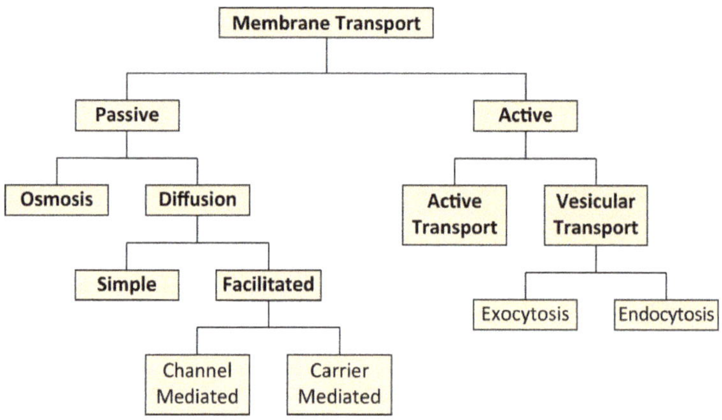

FIG. 7.6 Different types of cell membrane transport—rule of two

Diffusion/Simple Diffusion

Simple diffusion is the random movement of molecules in gases or liquids, from areas of high concentration to areas of low concentration, until equilibrium is reached. It is a passive process that does not require energy (ATP).

Molecules in gases, liquids, and solids have some movement or kinetic energy. This is also called Brownian movement. It is more pronounced in gases than liquids or solids. Think about water. Ice is the solid form of water and has a small amount of Brownian movement of the water molecules. When heat is added to ice it melts into water. With the added energy there is now a greater amount of random movement of the molecules. When additional heat is added to the water it turns into steam. Now there is a much greater amount of movement of the molecules.

This random movement (Brownian) is what creates the process of diffusion. It happens spontaneously without the need for additional energy.

✔ **CONCEPT!** Diffusion is the random movement of molecules from high to low concentration and does not require energy.

In physiology, diffusion occurs in liquids and gases in the body. For instance, oxygen moves from the alveolar sacs in the lungs into the blood stream by simple diffusion. Carbon dioxide moves from the blood stream into the alveolar sacs of the lungs by simple diffusion. At a tissue level, these gases move into and out of the cells by simple diffusion.

On a lighter note, a common example of simple diffusion everyone experiences is **flatus**, a fart! The release of intestinal gases results in a concentration of those gases just outside the anus. There is random movement of molecules in that "cloud" of gas. The molecules will then randomly spread out through simple diffusion attempting to reach a state of equilibrium. This has clinical significance after abdominal surgery when flatus indicates the patient's intestines are beginning to function again. The patient can then begin to return to eating, slowly with clear liquids and then a soft diet. So, flatus is important in medicine as well as a healthy and normal part of living!

✔ **CONCEPT!** A fun example of simple diffusion is a fart! So, next time you fart, you'll think about diffusion. Physiology is happening all the time and can be silly and fun!

Diffusion/Facilitated Diffusion

Simple diffusion takes place with or without the presence of a cell membrane. A fart or spraying air freshener after farting are examples of molecules diffusing without passing through a membrane. Facilitated diffusion takes place across a membrane.

Bridge back to the SPPLBL earlier in this chapter and review the structure of the SPPLBL or cell membrane. Do you remember what SPPLBL stands for?

Go online and put "animation of fluid mosaic model" in your browser. This will give you a variety of animations demonstrating the cell membrane in motion. It's a great way to refresh your understanding of the cell membrane and what is meant by the term *fluid mosaic* when describing the SPPLBL. It is essential for the novice student to have solid insight into the structure *and* function of the plasma membrane!

BRIDGE to Figure 7.2 on page 67 to review the cell membrane.

Notice the globular or integral protein in the SPPLBL. Those large protein molecules contained in the cell membrane are the site of the two types of facilitated diffusion, either channel-mediated or carrier-mediated diffusion.

Channel-mediated diffusion involves the movement of small particles across the cell membrane through integral proteins. Each channel is specific for only one type of ion, such as sodium or potassium. These channels can be either leak or gated channels. As you would expect, the names tell you how these channels function.

Leak channels remain open and allow constant movement of a specific ion., whereas gated channels are usually closed and will only open in response to a certain type of stimulus, such as a chemical agent, light, or an electrical stimulus. Gated channels only stay open for a fraction of a second before closing again.

Carrier-mediated diffusion facilitates the movement of small polar molecules such as simple sugars and amino acids. Polar molecules have an electrical charge, like water where the oxygen end is negative and the hydrogen end is positive.

✔ **CONCEPT!** Polar particles have an electrical charge; a sodium ion is positive, and a chloride ion is negative.

Remember all three types of diffusion, simple, carrier mediated, and channel mediated, are *passive* processes that *do not* require energy.

Osmosis

Another important type of cellular transport is osmosis. This is the diffusion of water across a semipermeable membrane toward a concentration of solute. Three things are required for osmosis to occur, a **solvent**, which in human physiology is always water, a **solute** such as salt or sugar, and an SPPLBL permeable to water but impermeable to the solute. The solute is dissolved in the solvent, creating a **solution.**

✔ **CONCEPT!** In human physiology, a solution consists of solutes (salt, sugar, etc.) dissolved or suspended in a solvent, specifically water.

✔ **CONCEPT!** Osmosis requires the presence of three things: solvent (H_2O), SPPLBL, and a solute (i.e., sugar or salt).

Figure 7.7 shows a simplistic diagram of osmosis. Notice the three things osmosis requires are present: water, SPPLBL, and solute. The membrane allows water to pass through it, but the solute is trapped on the right side of the diagram. Because the solute molecules take up space, there is a greater concentration of water on the left side than the right side of the beaker. Because the membrane allows the passage of water (but not solute) the water will *diffuse* from higher to lower areas of water concentration until equilibrium is reached. Because the membrane is permeable to water there is movement of water in both directions, but a *net movement* left to right. This results in more water in the right side than the left side of the beaker when equilibrium is reached.

Notice it is the water that is diffusing. Water is moving from higher to lower water concentration. Also notice the right side of the beaker has a much higher concentration of solute. So, the more solute, the more water will move toward that concentration of solute. The solute is figuratively sucking the water toward it.

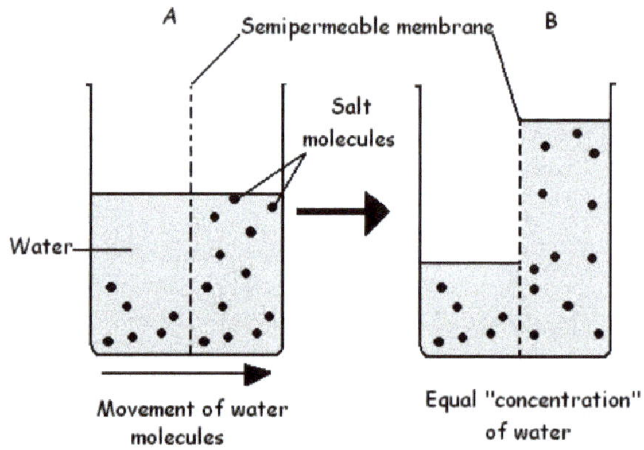

FIG. 7.7 Diffusion of water across a SPPLBL

✔ **CONCEPT!** **Solutes suck!** The higher the concentration of a solute on one side of a SPPLBL, the more water will move in that direction. So, it's easy to just remember that **solutes suck!**

Osmosis is an essential concept the novice A&P student must grasp. Remember, water makes up about two-thirds of human body mass. Water is also one of the two waste products of CR or cellular respiration. Obviously, the movement of water across cell membranes is vital to the body's ability to maintain homeostasis and health.

NOTES

Osmotic Pressure and Tonicity

As discussed in the previous section, a difference in solute concentration can (and usually does) exist between the cytosol (intracellular space) and the interstitial fluid outside of the cell. This difference exists because the SPPLBL prevents the solutes from moving from one space to the other. So, the solute is more concentrated on one side of the membrane than the other side. When a solute concentration exists, there is also a water concentration gradient because a solution with a high solute concentration will have a low water concentration and vice versa.

Osmolarity is the total concentration of all solutes in the solution and is related to tonicity. A solution with low osmolarity has fewer solute particles, while a solution with high osmolarity has more solute particles. When solutions of different osmolarities are separated by a membrane permeable to water, but not to solute, water will move from the side with lower osmolarity to the side with higher osmolarity.

The net movement of water into and out of cells is determined by the solute concentration inside the cell as compared with solute concentration outside the cell. Remember, solutes figuratively suck, so, the higher the solute concentration the greater the movement of water. This movement gradient is called osmotic pressure and it creates tonicity. **Tonicity** is the ability of a solution to change the volume or tone of a cell by adding or subtracting water through the process of osmosis.

In human physiology, we refer to three types of tonicity: hypotonic, isotonic, and hypertonic. Those terms refer to the concentration of solute (either salt or sugar) inside the cell as compared with the normal solute concentration in body tissues.

Human cells (like an RBC) have a concentration of 0.9 percent NaCl (salt) and a concentration of 5 percent sugar (glucose). These concentrations are referred to as being **isotonic**. The prefix "iso" means same. So, an isotonic solution contains the same concentration of salt and sugar as normal human cells. An isotonic solution of salt water is called **normal saline** meaning normal salt concentration. Many IV bags in a hospital have "Normal saline" printed on the bag, with a subtitle of "0.9 percent sodium chloride." Check out some IVs next time you're visiting or working in a hospital; just check in an appropriate and professional manner!

It's now easy to anticipate that a hypotonic solution will have less than 0.9 percent salt and/or less than 5 percent sugar. A hypertonic solution will have more than those "iso" amounts of sugar and salt. It's easy to understand why it's important to give patients isotonic solutions intravenously when you appreciate how solute concentration influences the movement of water. Remember, solutes suck, so the more solute (less water) in a solution the more osmotic pressure it will create.

To completely understand this concept, consider three beakers of water with different concentrations of solute, salt for instance.

- Hypotonic solution—distilled water with zero solute molecules
- Isotonic solution—containing 0.9 percent NaCl concentration
- Hypertonic solution—containing a concentration of salt that's greater than 0.9 percent

✔ **CONCEPT!** Figuratively, **solutes suck**, so water will move toward higher solute concentrations.

In the hypotonic solution, the concentration of salt is greater inside the cell than outside, so the net movement of water will be into the cell, causing it to swell and possible lyse (burst). In the isotonic solution, water will move into and out of the cell at an equal rate. Therefore, there'll be no net movement of water and the cell will remain the same size and shape. Remember, water is always moving into and out of cells, but solute concentration will affect **net movement**! That net movement is what creates changes in cell tonicity.

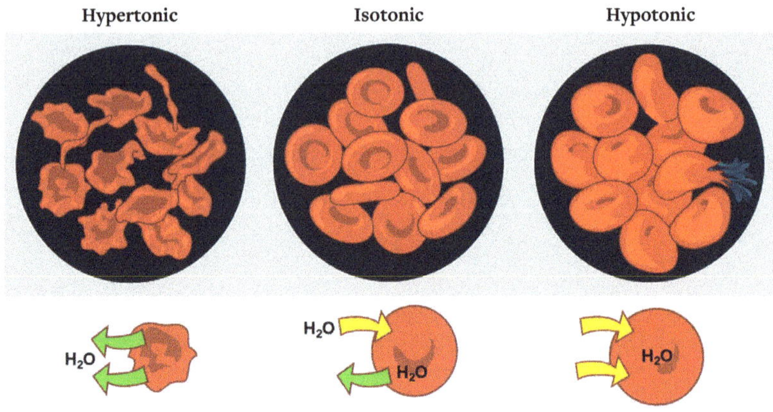

FIG. 7.8 Hypertonic, isotonic, and hypotonic solutions on RBCs

In a hypertonic solution, the osmotic pressure created by a greater concentration of salt in the solution will cause a net movement of water out of the cell, causing it to shrink or crenate. Figure 7.8 demonstrates tonicity and what would happen to a red blood cell (RBC) when placed in hypertonic, isotonic, and hypotonic solutions. It can also be helpful to go online and search for animations demonstrating tonicity.

NOTES

Active Transport (AT)

As Figure 7.6 shows, there are also active membrane transport processes. Active transport moves solutes **against** a concentration gradient. This requires the use of energy, either in the form of ATP produced by CR or the energy of a second molecule, which is diffusing down its concentration gradient, acting as a carrier molecule.

There are two types of active transport: primary and secondary (rule of two).

✔ **CONCEPT** Rule of two

- **Primary AT**—uses cellular energy (ATP) to move a substance against its concentration gradient

- **Secondary AT**—sometimes called cotransport. Energy is provided by the movement of a second substance down its concentration gradient, which pulls another particle with it. Think "piggyback" movement involving a carrier molecule.

Vesicular Transport

Vesicular transport is also called bulk transport. This process involves the transport of large substances or large amounts of smaller molecules. Cells form membrane-bound sacs in the cytoplasm filled with these large substances. These are many times proteins manufactured through the process of protein synthesis. The vesicle creation is a primary responsibility of the Golgi complex. This would be a good time to review cellular organelles, like the Golgi complex, as well as structures like the mitochondria that produce energy and ribosomes, which make proteins!

As we see so many times in A&P, vesicular transport has two subdivisions (rule of two).

BRIDGE to page 67 toward the front of this chapter to review cell structure and intracellular organelles.

✔ **CONCEPT** Rule of two: **Exocytosis** is a type of vesicular transport, moving particles from inside the cell to the extracellular space. **Endocytosis** is the opposite process, moving particles from outside the cell to the intracellular space.

- **Exocytosis**—the process of cells releasing substance from the intracellular space into the extracellular space, specifically the interstitial space
- **Endocytosis**—the process of cells ingesting substances from outside the cell to the inside of the cell

NOTES

Review Questions

- Why is the cell so important in understanding human A&P?
- What are the specific functions of each of the major cell organelles?
- What specific functions do the three cell surface modifications provide?
- What is an SPPLBL, and why is it important?
- What are the two functions of mitosis?
- What is simple diffusion?
- What is facilitated diffusion?

- What is osmosis?
- What three things must be present for osmosis to take place?
- What is tonicity, and why is it important?
- What's the difference between exocytosis and endocytosis?

Activities

- With colored pencils and/or crayons, draw and label a picture of the typical human cell, including the cell membrane, intracellular organelles, and nucleus. (Utilize the diagram of the cell in Figure 7.1 as a guide and/or go online and search for diagrams of a generic human cell, which will help you complete your drawing.)
- With colored pencils and/or crayons, draw a detailed picture of the SPPLBL or cell membrane. Include the phospholipid molecules (heads and tails), cholesterol, integral proteins, and surface molecules like glycoproteins and glycolipids. (Use the diagram in Figure 7.2 as a guide or go online and search for one.)
- Using colored pencils and/or crayons, draw a diagram of osmosis and explain it to someone else.
- Draw a diagram demonstrating hypotonic, isotonic, and hypertonic solutions and show the net movement of water into and out of the RBCs in each of the solutions. Figure 7.8 will be helpful.

FIGURE CREDITS

Churn to Learn!

Teach to Learn!

CHAPTER EIGHT

Histology

Tissues, Glands, and Membranes

TOES (TOPICS OF EMPHASIS)

- Definition of histology
- Four primary tissue types
- Naming of epithelium
- Specific types of epithelium with location and function
- Specific types of connective tissue
- Three types of muscle tissue
- Nervous tissue

Histology

This is the study ("ology" means "study of") of tissues. It is a detailed and complex science. But, the vast majority of students in an introductory A&P course are not going to be histologists. However, a conceptual understanding of histology is essential for the novice A&P student.

Primary Tissues

H_2O **CHAPTER 2** for details about the primary tissues

The human body is made up of four primary tissues. The H_2O sheet shows these four primary tissues in column two. Students are encouraged to repeat out loud the four letters of these tissues—E, C, M, N.

✔ **CONCEPT!** Repeat "ECMN" like a mantra. Read what follows with your H_2O sheet right in front of you. Concentrate on learning how understanding the H_2O sheet will allow you to gain real insight, eliminating the need to memorize without understanding.

- **Epithelium—lines and covers**
  - Any tissue that lines a body cavity or covers an organ or area (like the skin) is epithelium
  - Figure 8.1 shows examples of different types of epithelium

✔ **CONCEPT!** Understanding this simple concept: "E = L & C" can dramatically reduce the need to memorize dozens of specific types of tissue covering or lining organs. If the tissue lines or covers it is **epithelium**!

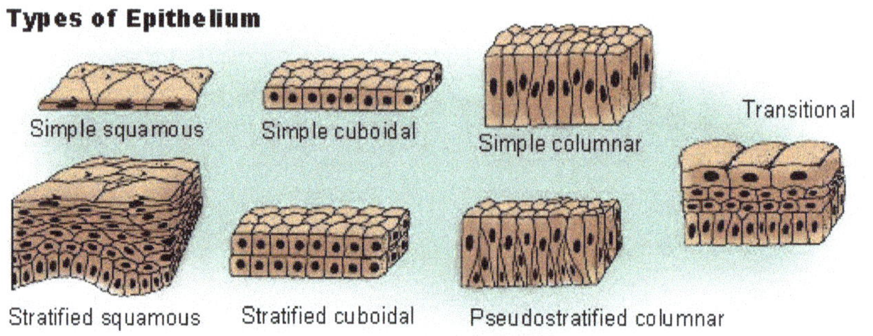

FIG. 8.1 Graphics showing different types of epithelium

- **Connective—connects** and **supports**. That's why it's called connective tissue! It connects stuff. For our purposes the term "supports" includes protection.

✔ **CONCEPT!** Tissue that connects stuff, fills space, protects, and supports is connective.

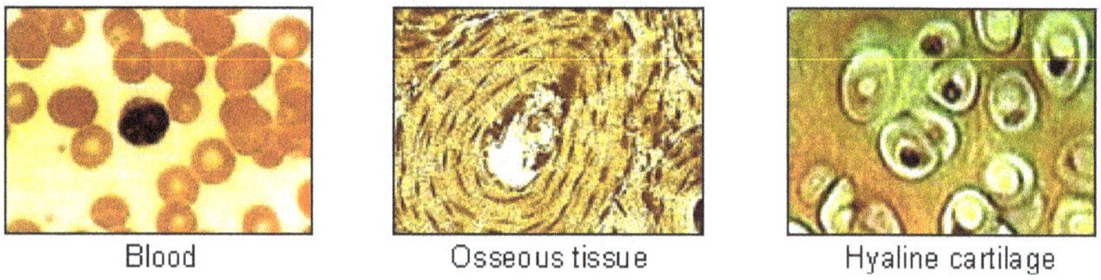

FIG. 8.2 Blood, bone, and hyaline cartilage. Examples of connective tissue.

  - Tendons connect skeletal muscle to bone, crossing joints, providing movement.
  - Ligaments connect bone to bone, supporting joints.
  - Ribs protect the lungs and heart.
  - Fat stores energy, protects and insulates.
  - Blood supports by transporting stuff like oxygen and nutrients.
  - Figure 8.2 shows examples of several types of connective tissue.
- **Muscle—provides movement**
  - Anytime stuff moves in the human body (with a rare exception) muscle is involved.

✔ **CONCEPT!** Muscle can do *only* one *active* thing and that is *contract* or shorten.

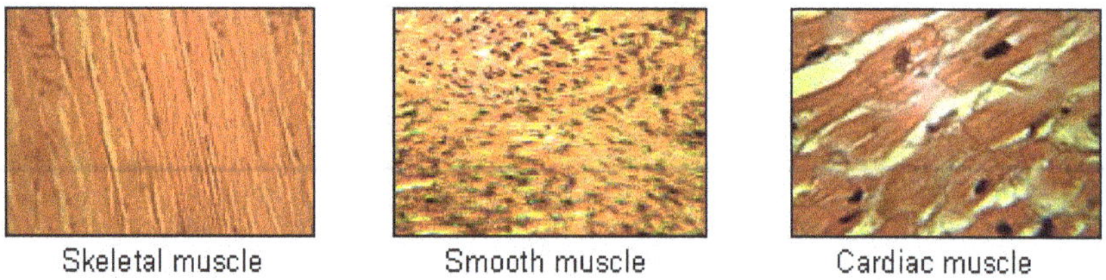

FIG. 8.3 Skeletal, visceral, and cardiac muscle

  - Three types of muscle:
    - Cardiac (only found in the heart)
    - Skeletal/striated/voluntary
    - Visceral/smooth/involuntary

- As occurs frequently in A&P, *you must know* the multiple names for this type of muscle! (multiple names occur a bunch in A&P!)

✔ **CONCEPT!** If the muscle is under voluntary control (you control it) then it must be skeletal muscle.

- Visceral/smooth/involuntary
  - Found in the "viscera" or the "guts" e.g. (lungs, blood vessels, digestive tract, etc.) of the body

---

✔ **CONCEPT!** Visceral muscle is not under voluntary control. It contracts involuntarily and automatically.

- If you can't control a muscle, then it is visceral (or cardiac) muscle.

- **Nervous—communication**
  - Nervous tissue is the body's communication tissue.
  - It gathers information from inside and outside the body and communicates that information to the brain and spinal cord for processing.
  - There are different structural and functional types of nerve cells that we'll cover when we get to the nervous system.

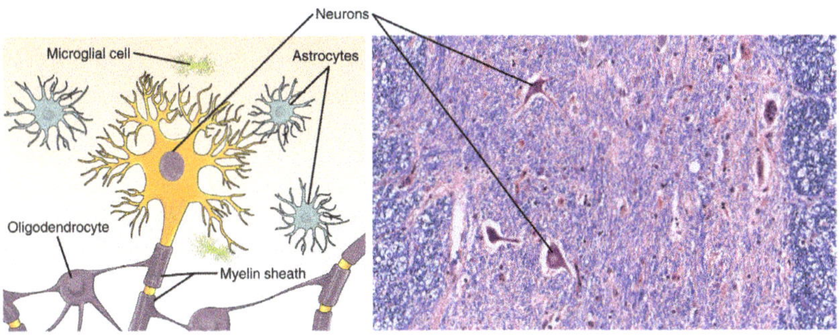

FIG. 8.4 Neurons and neuroglial cells of the nervous system

NOTES

Epithelial Tissue Types

Epithelial tissue lines and covers, covers and lines. There are many different types of epithelial tissue. We'll focus on nine specific types of epithelium.

✔ **CONCEPT!** Naming of epithelial tissue—layers and apical cell shape—layers/shape. Epithelial tissues are named by their thickness (layers) and the shape of the cells at the free or apical surface of the tissue.

✔ **CONCEPT!** Epithelial tissues are shaped the way they are because of what they do.

- Because epithelium lines and covers, it will always have a **free** or **apical surface** open to the outside (top layer of skin) or the lumen (opening) in a tube like the digestive tract or trachea.
- All epithelial tissue also has a **basement membrane** found at the bottom of the tissue. It is deep to the apical or free surface.
- Number of layers—first description in naming epithelium
  - Only one layer thick = **simple**
  - Two or more layers thick = **stratified**
- Shape of the cells at the free or apical surface of the epithelium
  - **Squamous** = flat—like a fried egg
  - **Cuboidal** = shaped like little cubes (Makes sense, huh?!)
  - **Columnar** = shaped like columns (Just what you'd expect!)
- Surface or functional modifications are added to the name of some epithelial tissues.
  - Keratinized (apical cells are dead) or nonkeratinized (apical cells are living)
  - Cilia
  - Microvilli
- Atypical naming for special epithelium tissue like **transitional** epithelium

- **Specific types of epithelial tissue** (you need to know the following):
  - **Simple squamous**—single layer of thin, flat cells facilitating movement of particles through the cells, as in the alveolar sacs of the lungs
  - **Simple cuboidal**—single layer of cube-shaped cells; found in kidney tubules
  - **Simple columnar**—single layer of columnar cells; found in majority of digestive tract
  - **Ciliated simple columnar**—single layer of ciliated columnar cells; found in the large bronchioles of lungs and fallopian (uterine) tubes
  - **Ciliated pseudostratified columnar**—single layer but appears multilayer because of the different location of the cell nuclei, so called falsely or pseudo stratified; may contain goblet cells that produce mucus; found in largest tubes of the lungs like the trachea
  - **Nonkeratinized stratified squamous**—dozens of layers of cells with flat living cells at the apical surface; found in oral cavity, esophagus, vagina
  - **Keratinized stratified squamous**—dozens of layers of cells with flat dead cells at the apical surface; found in the epidermis of skin

- **Stratified cuboidal**—two or more layers of cube-like cells; found in the ducts of most exocrine glands (go online and search "exocrine glands")
- **Transitional**—multiple layers with basement membrane and apical surface but with the unique ability to stretch; found in the urinary bladder

✔ **CONCEPT!** This is exactly what you'd expect if you apply FFF. The urinary bladder must stretch and expand to be able to store urine! So, transitional epithelium is structured the way it is because of what it does!

NOTES

Connective Tissue Types

Because of what it does, you might expect connective tissue to be the most diverse of the four primary tissue types. And you'd be right! See how applying **concepts** can facilitate your ability to gain insight into A&P rather than just memorizing hundreds of isolated facts you don't understand and will rapidly forget? If you grasp the concept, you're more likely to retain the information and be able to apply it in other courses, as well as in the real world of clinical health care.

✔ **CONCEPT** Understanding concepts helps you gain insight and minimize memorization

The novice A&P student will need to understand at least thirteen different types of connective tissue. As with epithelial tissue (basement membrane and apical surface), all connective tissues have similar characteristics—a matrix of ground substance with cells suspended in that matrix.

✔ **CONCEPT** Rule of two: two characteristics of epithelial tissue and two characteristics of connective tissue

- Three general types of connective tissue—each have two subcategories:
    - **Connective tissue proper**
        - Loose connective
        - Dense connective
    - **Supporting connective tissue**
        - Cartilage
        - Bone

- **Fluid connective tissue**
  - Blood
  - Lymph

- **Specific types of loose connective tissue proper**—fewer fibers, more ground substance (as you'd expect since it's called "loose")
  - **Areolar tissue**—found throughout the body, surrounds most organs and vessels; acts as the body's filler, much like those packing peanuts you use when packing something to ship or move
  - **Adipose tissue**—closely packed cells, stores energy, protects, and insulates; found throughout the body, around organs and in the subcutaneous layer deep to the skin
  - **Reticular tissue**—viscous (go online and search for the definition) ground substance, scattered fibers and leukocytes; provides support for lymphatic tissue; found in lymph nodes, spleen, and red bone marrow

- **Specific types of dense connective tissue**—more fibers and less ground substance
  - **Dense regular tissue**—densely packed parallel arrangements of collagen fibers; provides maximum strength in one direction or plane; found in tendons and ligaments
  - **Dense irregular tissue**—collagen fibers arranged randomly, clumped together; provides strength in many directions; found in dermis of skin, periosteum of bone, and joint capsules
  - **Elastic connective tissue**—composed mostly of elastic fibers allowing for stretching and recoil; found in large arteries, like the aorta, and in the trachea and vocal cords

- **Specific types of supporting connective tissue (cartilage)**—strong, durable framework that supports and protects soft tissues (viscera)
  - **Hyaline cartilage** (matrix appears glass like)—loosely arranged chondrocytes in lacunae; most common type of cartilage; found in larynx, costal cartilage, articular cartilage of joints, tip of the nose
  - **Fibrocartilage** (weight-bearing cartilage)—numerous parallel collagen fibers; found *only* in intervertebral discs, symphysis pubis, and menisci of the knee
  - **Elastic cartilage** (many elastic fibers in a web-like mesh)—maintains shape while being flexible; found *only* in the epiglottis and external ear

- **Specific types of supporting connective tissue (bone)**—calcified extracellular matrix surrounding osteocytes trapped in lacunae
  - **Compact bone** (arranged in osteons)—makes up shafts of long bones
  - **Spongy bone** (meshwork pattern)—softer than compact bone

- **Specific types of fluid connective tissue—blood and lymph**
  - **Blood**—contains formed elements (FEs—blood cells) suspended in a liquid matrix; plasma
  - **Lymph**—derived from blood, absorbed interstitial (between cells) fluid plus fat molecules absorbed into the blood stream in the small intestine, eventually emptied into the subclavian veins.

BRIDGE to Chapters 14 and 15 for more details about blood and lymph

NOTES

Muscle Tissue Types

Muscle tissue is composed of specialized cells that can contract or shorten when stimulated.

✔ **CONCEPT!** Muscle allows movement of the body and movement of stuff within the body. There are three specific types of muscle tissue structured the way they are because of what they do.

- **Cardiac muscle**—short, striated (lightly striped), branching pattern with intercalated discs caused by gap junctions between cells; **involuntary** control; only found in the heart
- **Striated, skeletal, or voluntary muscle**—long, striated (deeply striped), nonbranching with many nuclei and with fibers (cells) that are parallel; **voluntary** control
- **Visceral, smooth, or involuntary muscle**—nonstriated (smooth) fibers that are short and fusiform in shape with one central nucleus; **involuntary** control; found in walls of hollow organs such as the stomach and other viscera like blood vessels

✔ **CONCEPT!** Any structure an individual can voluntarily control will contain skeletal/striated or voluntary muscle. Students can minimize memorization by being able to think! For example, our diaphragm (the main muscle of breathing) must be made of skeletal muscle because we can control our breathing. On the other hand, the uterus in the female must contain smooth or visceral muscle because a woman cannot voluntarily contract her uterus. A&P is more fun when students can figure things out!

Nervous Tissue

Nervous tissue exists in the brain, spinal cord, and nerves throughout the body. Nervous tissue is made up of neurons that are excitable and transmit electrical impulses, as well as neuron support cells called glia. There are roughly ten times as many glial cells as neurons. Glial cells do not transmit nerve impulses; only neurons do that.

Bridge to Chapter 12, page 156 to gain additional insight into the six neuroglial support cells.

BRIDGE to Chapter 12, page 156 for more information about glial cells

NOTES

Time to Churn to Learn

Students *must* spend time studying pictures or microscope slides of different tissues to become confident in their understanding of histology. The internet provides many open public sites showing excellent representations of the primary tissues and tissue subtypes.

Simply put "human histology" into a browser and you'll find a variety of resources that will help you gain insight into human tissues. A good site is "Histology-World." "Wikimedia Commons" provides a wide selection of images of human tissues. So, spend a little time surfing until you find a site or a couple sites that meet your needs. Searching histological images on your own helps to solidify your learning. Remember, the ones doing the most work are the ones doing the most learning!

Although a basic, conceptual understanding of histology is essential for the novice A&P student, it does not require any significant level of histological expertise. Those higher levels of understanding involve the upper floors of the four-story structure we're building. For the time being, we're staying in the basement building a solid foundation.

✔ **CONCEPT!** Review how epithelium is named. First, determine the number of layers, either simple (one layer) or striated (more than one layer). Then what is the shape of the cells at the apical or free surface, flat (squamous), cuboidal, or columnar. Remember there are also unique types of epithelium which don't follow this common naming pattern, like transitional epithelium (urinary bladder).

Review Questions

- What is histology?
- What are the four primary tissue types? (Utilize the H_2O sheet)
- How are epithelial tissues named?
- What are the three primary connective tissue types?
- What are neuroglia cells?
- What does epithelial tissue do? (See the H_2O sheet)
- What does connective tissue do? (See the H_2O sheet)
- What is the *only* active thing muscle tissue does?
- What are the five Cs of the nervous tissue? (See Chapter 2)

Activities and Coloring

Get out those colored pencils and/or crayons and some blank sheets of paper. Using the images you've found online, draw the following 12 tissues. Remember you're not striving to become an artist, but the effort/struggle to draw creates more new synapses (if you don't know what a synapse is, look it up!) and deepens your understanding of human tissues.

- Simple squamous epithelium
- Keratinized stratified squamous epithelium
- Transitional epithelium
- Ciliated pseudostratified columnar epithelium
- Compact bone
- Dense regular connective tissue
- Hyaline cartilage
- Elastic connective tissue
- Blood
- Cardiac muscle
- Striated muscle
- Multipolar neuron

NOTES

FIGURE CREDITS

Figure 8.1: Source: https://commons.wikimedia.org/wiki/File:Illu_epithelium.jpg

Figure 8.2: Source: https://commons.wikimedia.org/wiki/File:Illu_connective_tissues_2.jpg

Figure 8.3: Source: https://commons.wikimedia.org/wiki/File:Illu_muscle_tissues.jpg

Figure 8.4: Copyright © OpenStax College (CC BY 3.0) at https://commons.wikimedia.org/wiki/File:416_Nervous_Tissue-new.jpg.

Churn to Learn!

Teach to Learn!

CHAPTER NINE

The Skin

Integumentary System

TOES (TOPICS OF EMPHASIS)

- Skin = integumentary system = cutaneous membrane
- Structure of integumentary system—classic x-sectional image
- Conceptual functions of skin
- Layers of the epidermis
- Accessory structures, receptors and glands (sweat, oil)

Skin or Integumentary System

Skin is the system that covers our bodies. It is the *largest* organ of the entire body but *not* the largest organ *in* the body. The liver is the largest organ inside the body. Skin is also the only system considered to be an organ. The integumentary system is made up of the skin and its derivatives such as nails, hair, and glands.

Figure 9.2 is a classic picture of the cross-section of human skin. It is important for you to spend time reviewing this diagram. Let's examine some of the specifics of skin.

✔ **CONCEPT!** The skin is structured the way it is because of what it does: **contains and protects (H_2O sheet)**.

Layers

Skin is layered, as most of us have discovered when we develop a blister or experience a burn or other type of skin injury. Skin is composed of two basic layers.

(Rule of two popping up *again*!)

Skin Flow Chart

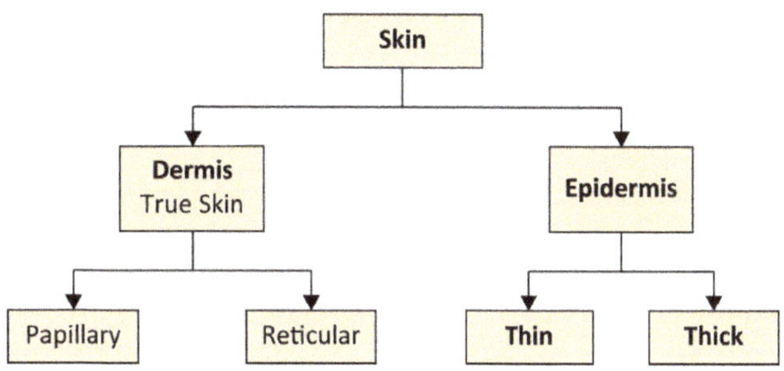

FIG. 9.1 Skin Flow Chart—rule of two

- **Dermis**—known as the true skin; it is the thickest layer of the skin and is filled with a variety of support and accessory structures
  - Blood vessels
  - Nervous receptors
    - Tactile or Meissner's corpuscles—soft touch
    - Lamellar or Pacinian corpuscles—pressure or heavy touch
    - Free nerve endings—pain and temperature

✔ **CONCEPT!** Tactile and lamellar corpuscles are located where they are and shaped the way they are because of what they do! Tactile are superficial, providing light touch, while lamellar are deep, providing pressure or deep touch.

  - Hair follicles
    - Hair shaft
    - Arrector pili muscle
  - Sebaceous glands
  - Sweat glands
  - Oil or sebaceous glands
- **Epidermis**—the layer above the dermis

The subcutaneous layer—also known as the hypodermis—is not considered as a part of the skin, since it is mainly adipose (fat) tissue. In Figure 9.2 notice the dermal papilla—ridges that secure the epidermis to the dermis.

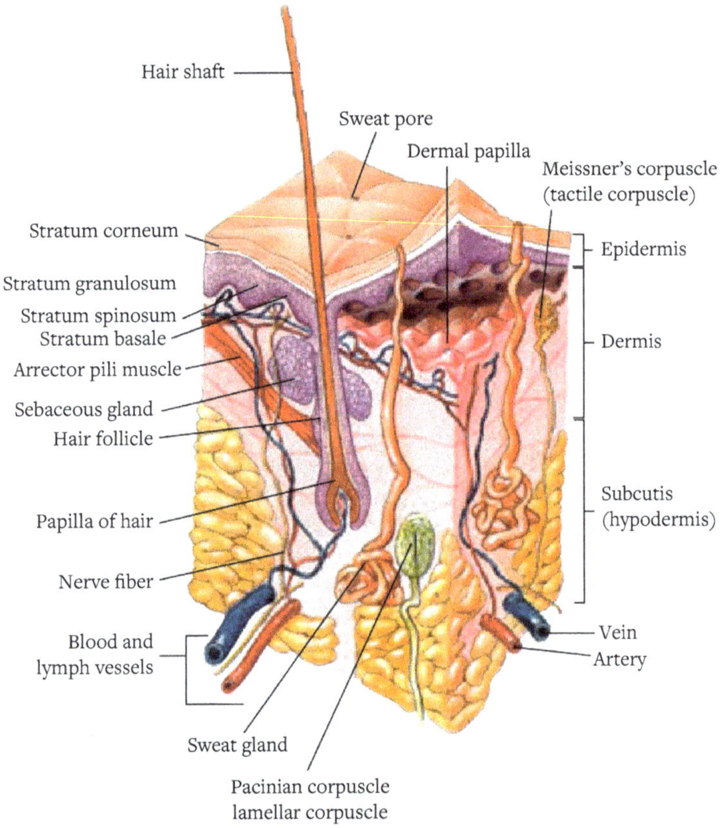

FIG. 9.2 Cross-section of human skin, showing major structures

NOTES

Layers of the Epidermis

A little deductive logic allows you to know with confidence that the epidermis is a type of epithelium. It is covering the dermis at the surface of the skin. Depending on the location of the skin, the epidermis has either four or five layers. The epidermis is structured as it is because of what it *does* (FFF)!

- **Thin skin**—four layers; found throughout the entire body *except* palms and soles
- **Thick skin**; five layers found *only* on the palms of the hands and soles of the feet
- Keratinized, stratified squamous epithelium
- Basal layer with a free, apical surface—as with all epithelial tissues
- Layers from deep to superficial:
  - Stratium basale—only layer where mitosis takes place
  - Stratium spinousum
  - Stratum granulosum
  - Stratum lucidum (only on the palms and soles)
  - Stratum corneum—dead scaly keratinized cells (twenty to thirty layers thick)
- Avascular—no blood vessels
- Melanocytes in the basilar layer produce melanin—pigment
- Two specialty cells
  - Epidermal dendritic cells or Langerhans cells—provide defensive function
  - Merkel cells—nerve receptor cells sensitive to pressure and position

Layers of the Dermis

The dermis is deep to the epidermis and is referred to as the **true skin**. It is composed of connective tissue proper and contains a large amount of collagen fibers. It also contains a rich supply of blood vessels as well as sweat glands, sebaceous glands, hair follicles, sensory nerve endings and receptors, and arrector pili muscles. The arrector pili muscles cause goose bumps when they contract.

✔ **CONCEPT!** Arrector pili muscles are what specific type of muscle tissue? Do you have voluntary control over "goose bumps"? Think about it!

The dermis is composed of two layers:

- **Papillary layer**—this is the more superficial layer of the dermis and is located directly under (deep) to the epidermis.
  - Composed mostly of loose areolar connective tissue
  - Contains capillaries
  - It derives its name from the **dermal papillae**
  - Papilla = nipple
  - Dermal papillae interdigitate with deep projections on the epidermis called epidermal ridges

- Papillae and ridges fit together much like two sets of egg crates stacked on top of each other. This structure locks the dermis and epidermis together.
- **Reticular layer**—forms the deeper and major portion of the dermis, extending down to the subcutaneous layer
  - Primarily consists of dense irregular connective tissue
  - Large bundles of collagen fibers woven into dense networks
  - Reticular = network of fibers
  - Contains most of the accessory structures of skin such as hair follicles and glands
- **Subcutaneous layer**—is not usually considered a part of the integumentary system
  - Deep to the dermis
  - Composed mostly of areolar and adipose connective tissue

FIG. 9.3 Egg cartons interlocking like the dermis and epidermis

NOTES

Integumentary structures derived from epidermis—hair, nails, and exocrine (with ducts) glands—are derived from the epidermal epithelium.

✔ **CONCEPT!** ALL glandular tissue in the body is derived from the epidermis.

- Also known as **epidermal appendages or derivatives**
- **Nails**—composed primarily of dead, keratinized epithelial cells
  - Parts of nail include nail bed, root, body, and plate
  - Eponychium = cuticle
  - Changes in appearance of nails can indicate a variety of internal diseases or dysfunctions
- **Hair**—found almost everywhere on the body except the palms and soles
  - Structure includes hair bulb, papilla, root, and shaft
  - Arrector pili muscles attaches to hair follicle; smooth muscle; goose bumps

- **Sweat glands**—two types:
  - **Eccrine or merocrine**—located throughout the body, including palms and soles
    - Important in helping to maintain homeostatic balance of body temperature control
    - Secrete through ducts (exocrine gland) to the surface of the epidermis
  - **Apocrine**—located only in a few areas:
    - Beard of adult male
    - Areolar area around nipple of adult female
    - Axilla in both male and female
    - Genitorectal area of both sexes
    - Odorous secretions produced
    - Sweat is secreted into the hair follicle and not directly to the surface of the skin as do eccrine glands
    - Sexual attraction through pheromones is proposed function
  - **Sebaceous glands**—produce an oily, waxy substance
    - Sebum
    - Oil lubricates and softens hair
    - Sebum (oil) is secreted into hair follicle directly on to the hair shaft
- **Burns** to the skin are usually caused by thermal exposure but can be chemical or from radiation.
- Three types of burns:
  - **First degree** (partial thickness of the skin)—only involves the epidermis
    - Pain
    - Redness
    - No blisters
  - **Second degree** (partial thickness)—entire epidermis and part of the dermis
    - Pain
    - Redness
    - Blisters
  - **Third degree** (full thickness)—involves epidermis, dermis, and subcutaneous layer
    - Painless—nerves in the skin have been destroyed
    - Black or gray in appearance
    - Hypovolemic shock (loss of ability to contain body fluids)
    - Infection

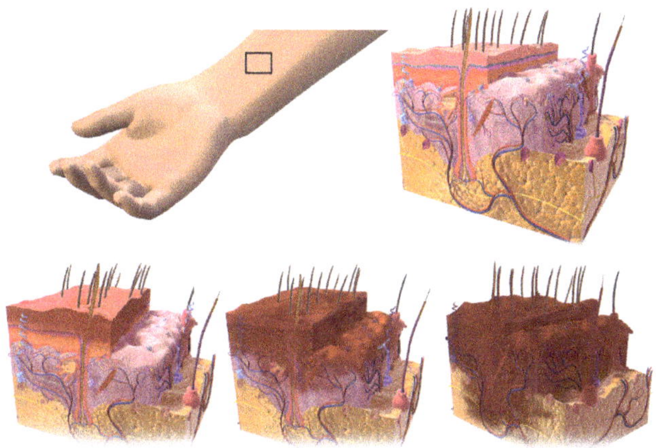

FIG. 9.4 Partial and full-thickness burns

✔ **CONCEPT!** Look at the fourth column, systems, in the H_2O sheet. Integumentary has C&P as an abbreviation for its function.

- **Containment**—holds the body's fluids (mostly water) inside; when burns damage skin the body loses fluids, causing dehydration leading to hypovolemic shock. So, immediate fluid replacement and stabilization of fluid balance is *essential* when initially treating significant body burns.
- **Protection**—once a patient is fluid balanced, preventing infection is the next significant concern because the loss of skin makes it difficult for the body to defend against infection.

✔ **CONCEPT!** **Rule of nines**—a method used by emergency medical personnel to estimate the area of the body burned. It divides the body surface into areas representing 9 percent, or multiples of 9 percent of body area.

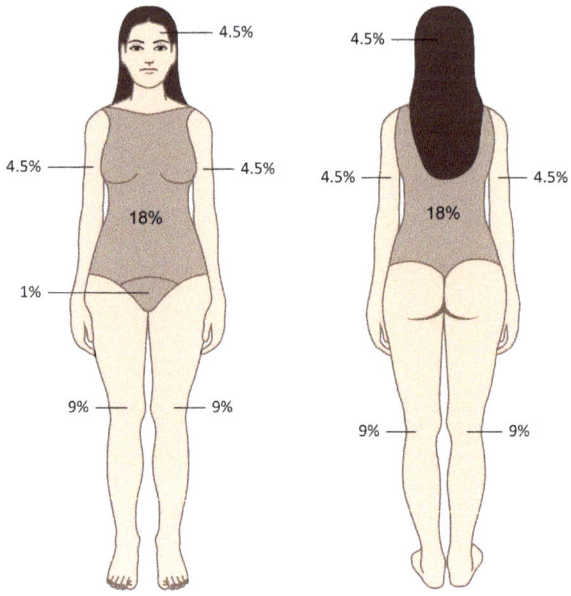

FIG. 9.5 Rule of nines allowing estimate of area of skin burned

Skin cancer—three types and three levels of severity:

- **Basal cell carcinoma**—originates from the stratum basale
  - Related to sun exposure, especially severe sunburn in the past
  - Most common type of skin cancer
  - Least dangerous skin cancer
  - Noninvasive—will grow but generally does not invade surrounding tissues
  - Nonmetastatic—usually does not spread to other areas of the body
- **Squamous cell carcinoma**—originates from keratinocytes in the stratum spinosum
  - Related to sun exposure
  - Second most common type of skin cancer
  - Invasive—will invade surrounding tissues
  - May metastasize but usually is non-metastatic

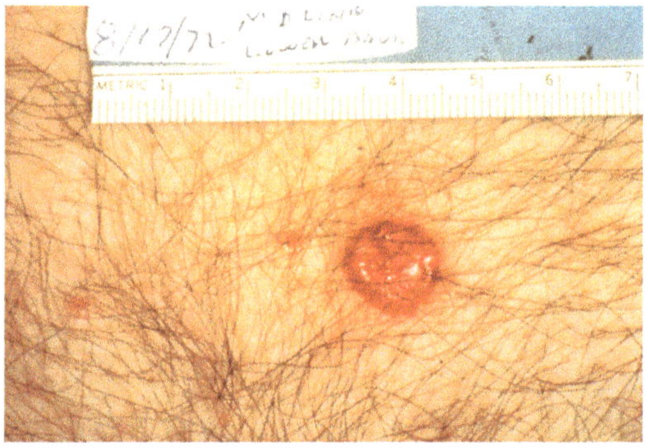

FIG. 9.6 Typical basal cell carcinoma—most common skin cancer

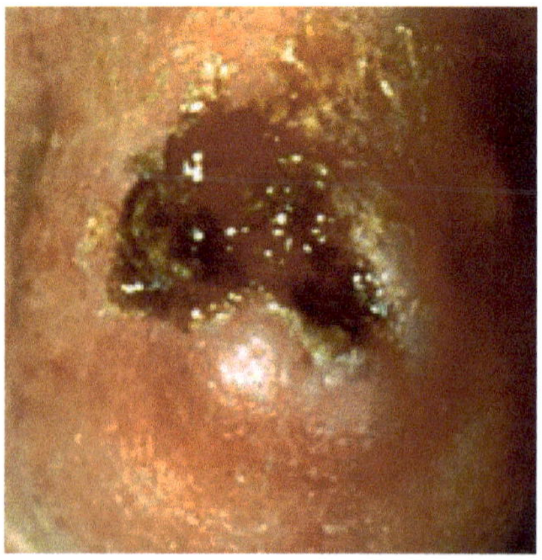

FIG. 9.7 Squamous cell carcinoma

- **Malignant melanoma**—arises from melanocytes in the basilar layer of the skin
  - Deadly type of cancer—serious in nature
  - Aggressive growth
  - Aggressive metastasis
  - Difficult to treat once it has spread
  - May be related to sun exposure
  - Can occur in areas with little or no sun exposure
    - Soles of feet
    - Genital area
  - ABCD rule for recognizing malignant melanoma: Asymmetry, Border irregularity, Color, Diameter

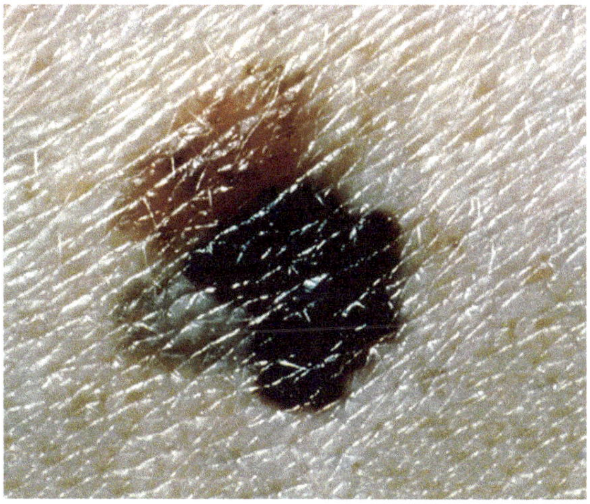

FIG. 9.8 Malignant melanoma—most deadly type of skin cancer

NOTES

Review Questions

- What is the "true" skin?
- How does the integumentary system exemplify the concept of FFF?
- What are the three types of burns?
- What is the "rule of nines," and how is it utilized?
- What are the three types of skin cancer?
- Which type of skin cancer is the most common?
- Which type of skin cancer is the most dangerous?

Activities and Coloring

Get out those colored pencils and/or crayons again and on a full sheet of paper draw out a cross-sectional diagram of human skin, as seen in Figure 9.2. Utilize this diagram the first time you draw the skin, but then repeat your drawing from scratch at least one more time from memory. Remember, repetition is an essential key to learning!

FIGURE CREDITS

Figure 9.2: Source: https://commons.wikimedia.org/wiki/File:Skin.png.
Figure 9.3: Copyright © 2015 Depositphotos/dadalia.
Figure 9.4: Copyright © by BruceBlaus (CC BY-SA 4.0) at https://commons.wikimedia.org/wiki/File:1st,_2nd,_and_3rd_degree_burns.jpg.
Figure 9.6: Source: https://commons.wikimedia.org/wiki/File:Basal_cell_carcinoma_%282%29.jpg.
Figure 9.7: Source: https://commons.wikimedia.org/wiki/File:Squamous_cell_carcinoma_crop.jpg.
Figure 9.8: Source: https://commons.wikimedia.org/wiki/File:Melanoma_(4).jpg.

Churn to Learn!

Teach to Learn!

UNIT II

SUPPORT AND MOVEMENT

CHAPTER TEN

Skeleton and Bone Physiology

Calcium Homeostasis

TOES (TOPICS OF EMPHASIS)

- Anatomy of bone
- Conceptual classification of bone
- Axial skeleton
- Appendicular skeleton
- Joints
- Osteon in compact bone
- Bone remodeling—osteoblasts and osteoclasts
- Parathyroid hormone
- Calcitonin
- Negative feedback loops—calcium homeostasis

Bones

The adult skeleton is made up of a total of 206 different bones. Don't memorize that number, but it's a fun piece of trivia, and it could be a Jeopardy answer! And now that you've been told "not" to memorize it, you'll likely remember that number 206 for years! Just like when someone says "don't think about an elephant in the room" and then that's all you can see in the room, an elephant! Interesting how the human mind works.

Bone is the major organ of the skeletal system. Bone provides structural support as well as a framework for the body. The skeleton provides attachment sites for the insertion of the tendons of skeletal muscles. In conjunction with joints, this allows movement of the body.

Bone also is the site of the production of all the blood cells: red, white, and platelets. And it is a storage site for essential minerals such as calcium and phosphate.

Calcium is essential for bone strength, the normal function of all three types of muscle (cardiac, skeletal, and visceral), normal function of synapses necessary for nerve transmission, and normal blood clotting. So, obviously, the homeostasis of calcium is essential to good health.

Bone Classification

Bones appear in various shapes and sizes depending on their function. Hopefully that sounds familiar to you. Remember the basic concept for all of A&P!

✔ **CONCEPT!** Anatomical structures are shaped the way they are because of what they do.

The **four classifications** of bone in the human are as follows:

- **Long bones**—called long because they are long: greater in length than in width.
  - Femur
  - Humerus
  - Tibia
  - Radius
  - Clavicle

✔ **CONCEPT!** Anatomy (and physiology) are very logical sciences, and many times the name of a structure tells you about that structure. For example, a long bone is long, and the left A-V valve of the heart is located between the left atrium and left ventricle of the heart. Strive to understand the concept of common sense. It will make learning much easier and more fun!

- **Short bones**—called short because they are short, with length nearly equal to width.
  - Carpals (wrist bones)
  - Tarsals (ankle bones)
  - Patella or kneecap or other sesamoid bones (stop a minute and search online for a definition of sesamoid bone)
- **Flat bones**—called flat because they are flat. They have flat, thin surfaces that provide extensive area for muscle attachment as well as protection of underlying soft tissues.
  - Bones of the skull
  - Scapula—shoulder blade
  - Sternum—breastbone
  - Ribs
- **Irregular bones**—called irregular because (you guessed it) they are irregular in shape because of what they do. Are you starting to see how FFF applies to A&P? These bones have complex shapes because of what they do!
  - Coxae or hip bones
  - Vertebrae in the spinal column
  - Bones of the skull
    - Sphenoid
    - Ethmoid
    - Zygomatic

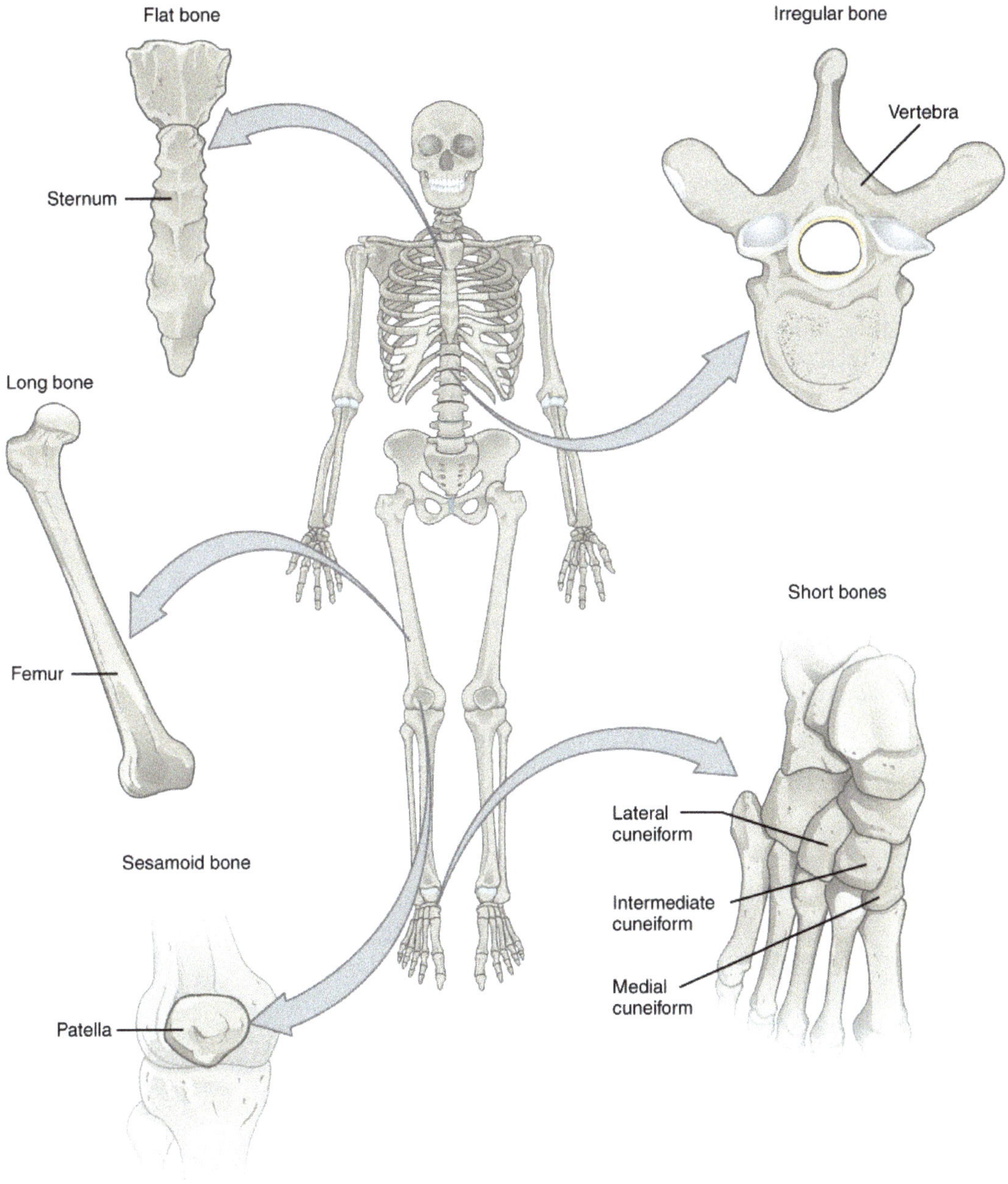

FIG. 10.1 Long, short, flat, irregular, and sesamoid (a bone contained in a muscle tendon—e.g., the patella)

Anatomy of Long Bones

Long bones are the most common of the four bone shapes in the body and so serve as a good general model for bone structure. These long bones are made mostly of compact bone with areas of spongy bone toward each end of the bone.

As Figure 10.2 demonstrates, you need to know the following features, areas, or structures of a typical long bone:

- **Diaphysis**—shaft of the bone
- **Medullary cavity**—cylindrical cavity in the center of the bone; contains bone marrow
- **Epiphysis**—end of the long bone
    - Proximal epiphysis—end of bone closest to the trunk of the body
    - Distal epiphysis—end of bone farthest from the trunk
- **Articular cartilage**—thin layer of hyaline cartilage covering joint surface of bone
- **Metaphysis**—area between the diaphysis and epiphysis (contains epiphyseal line)
- **Epiphyseal line in adults**—remnant of the growth plate (epiphyseal plate)
- **Epiphyseal plate in children and adolescents**—area where bone growth occurs
- **Spongy bone**—located at the epiphysis, both proximal and distal ends of bone
    - Inside a thin layer of compact bone
    - Provides strength in many directions
- **Nutrient foramen**—opening for blood vessels
    - Bone is vascular, as are all other (with an exception or two) tissues in the body.
    - Bone is constantly being absorbed and laid down. It is dynamic and changing.
- **Periosteum**—tough sheath covering the entire outside of the bone except articular surfaces
    - Made of dense irregular connective tissue
    - Attachment site for ligaments and tendons
- **Endosteum**—a thin layer of connective tissue lining the inner surface of the medullary cavity of long bones.

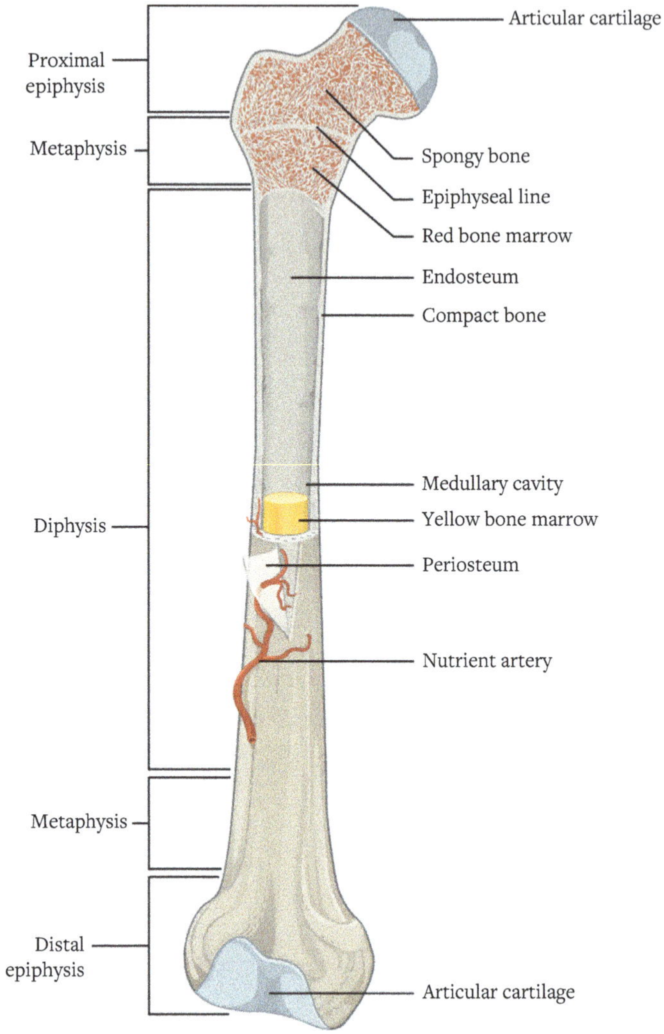

FIG. 10.2 Typical long bone structure (e.g., the femur)

The structure of short, flat, and irregular bones is different than long bones. The external surface in bones other than long bones is compact bone. The interior of the bone is made entirely of spongy bone. There is no medullary cavity.

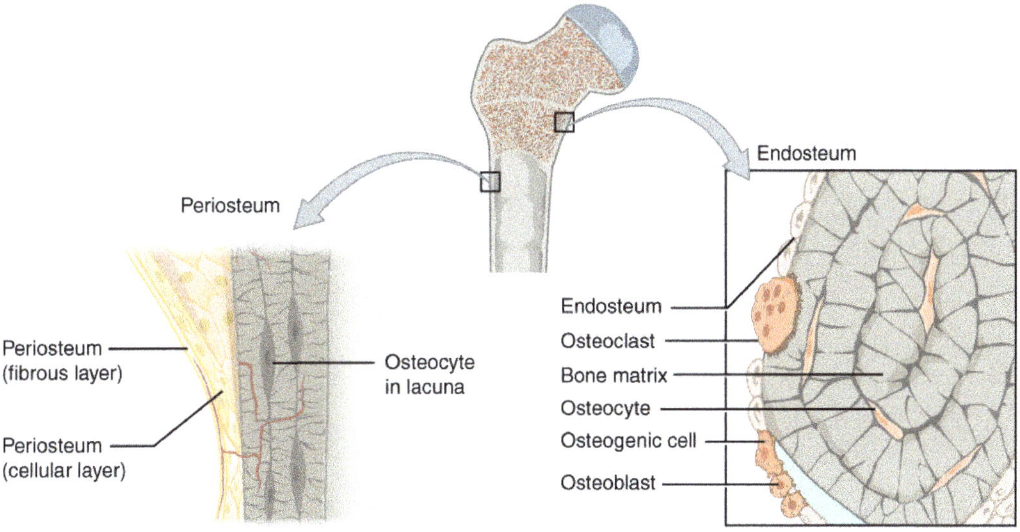

FIG. 10.3 Periosteum covering the outside of bone; endosteum lining the inside of the bone and marrow cavity

NOTES

Skeleton

The adult skeleton has 206 bones organized into two general divisions: the axial and the appendicular skeleton. Remember the rule of two; this is another example!

- **Axial skeleton**—made up of only four areas of bone
  - Skull including the mandible (jaw bone)
  - Vertebral column
  - Thoracic cage—including the sternum
  - Hyoid bone—the *only* bone in the body that doesn't articulate with (join) another bone
- **Appendicular skeleton**—all other bones that are not part of the axial skeleton
  - Clavicle

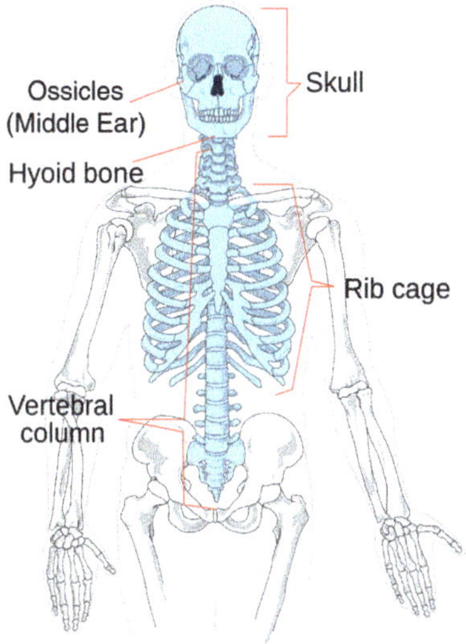

FIG. 10.4 Skull, vertebral column, thoracic cage, and hyoid. Can include the bones of the middle ear.

- Humerus
- Femur
- Tibia

✔ **CONCEPT!** Remember the bones that make up the axial skeleton (four general regions, skull, vertebral column, and the thoracic cage plus the hyoid bone). Memorize them, and then any other bone must be part of the appendicular skeleton. Once again, understanding concepts will minimize memorization, not eliminate it, but significantly reduce the amount of sheer memorization.

Remember, *SS A&P* can be utilized independently, as a standalone guide, or in conjunction with images and drawings from a traditional A&P reference textbook, e-text, or a variety of online sources. Students can use *SS A&P* to help guide them as they identify the essential/must-know structures of the skeleton and skeletal muscles. This applies to other organs and organ systems as well and will help students gain A&P insight. Students can also benefit from models and anatomical specimens available in an A&P lab or tutoring room.

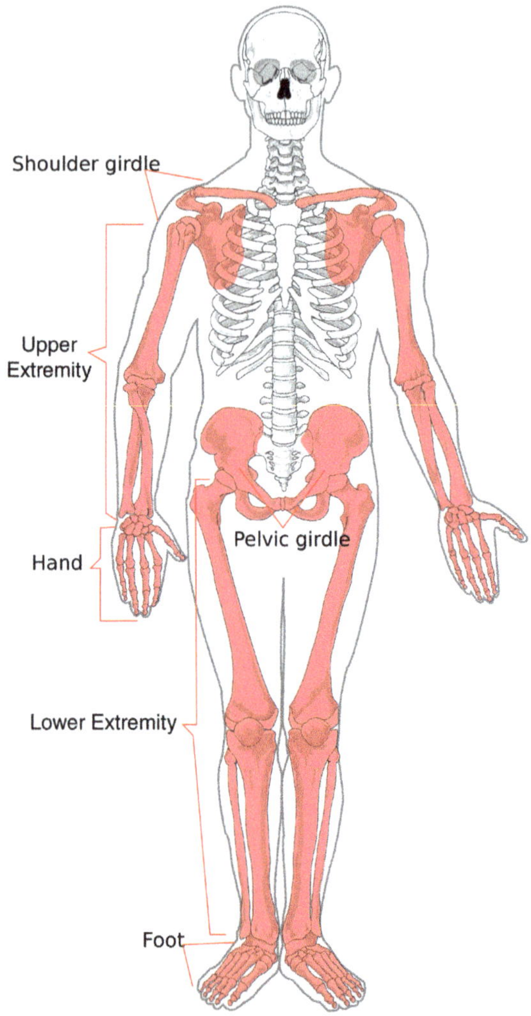

FIG. 10.5 All bones except those of the axial skeleton, including upper and lower extremities

Bone Markings

Bones have distinctive markings that characterize each bone. Projections (bumps and ridges) mark the surface where muscles, tendons, and ligaments attach. Depressions, grooves, and openings indicate sites where blood vessels and nerves travel.

✔ **CONCEPT!** If there is a hole (any kind of a hole) in a bone it is because stuff runs through that hole. The hole is there because of what it does, which is provide an opening for stuff like blood vessels and nerves. So, if there's a hole in a bone, you know something runs through that hole (with an exception or two!).

It is essential that you do the hard work to look up the following markings, openings, and bones of both the axial and appendicular skeleton. I know it's a ton of work, but you have to spend the time if you want to gain the insight and understanding you'll need to be successful. So, get to work. **Remember, the ones doing the most work are the ones doing the most learning!**

Bone markings summary (you need to know the following):

- **Articulating surfaces**—these are surfaces that create joints
  - Condyle—large, smooth, rounded surface

  - Facet—small, flat, shallow surface
  - Head—prominent, rounded epiphysis
  - Trochlea—smooth, grooved, pulley-like process
- **Depressions**—indentations, like a ditch in bone
  - Alveolus—deep socket in the upper and lower jaw; think teeth
  - Fossa—shallow depression
  - Sulcus—narrow groove (we'll see sulci [plural] in the brain as well)
- **Projections**—areas where ligaments and tendons attach, causing bone to thicken
  - Crest—narrow, ridge-like projection
  - Epicondyle—projection located above or adjacent to a condyle
  - Line—just what it sounds like: a low ridge
  - Process—a variety of bony prominences
  - Ramus—angular projection relative to the rest of a structure (jaw)
  - Spine—pointed, slender process
  - Trochanter—massive, rough process found only on the femur
  - Tubercle—small, round projection
  - Tuberosity—large, rough projection
- **Openings and spaces**—stuff runs through them!
  - Canal—passageway through bone, such as the optic canal
  - Fissure—narrow, slit-like opening
  - Foramen—one of the most frequently used terms for an opening; rounded passageway through bone
  - Meatus—another name for a passageway or opening
  - Sinus—cavity or hollow space in bone (paranasal sinuses)

Figure 10.6 shows a photograph of the base of a human skull with some of the most important and visible foramina labeled.

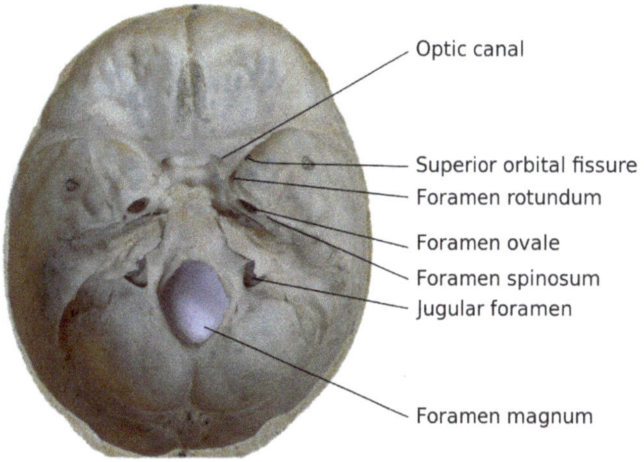

FIG. 10.6 Multiple openings in the base of the skull.

NOTES

Axial Skeleton—Skull

The skull is made up of 22 bones and is the most complex bone in the body. This complexity is caused by what the skull does! It houses and protects the brain and related structures and organs, such as the pituitary gland and the eyes.

You can benefit from a traditional text, online resources, anatomical models, and an effective instructor or tutor. *SS A&P* shows you the specific stuff you need to know to build a solid A&P foundation. Utilizing a variety of resources will supplement the foundation *SS A&P* allows you to build.

Must-know foundational information about the skull and hyoid bone:

- **Major bones of the skull**
  - Frontal
  - Parietal
  - Occipital
  - Temporal
  - Sphenoid
  - Ethmoid
  - Zygomatic
  - Lacrimal
  - Vomer
  - Nasal concha
  - Palatine
  - Maxilla
  - Mandible

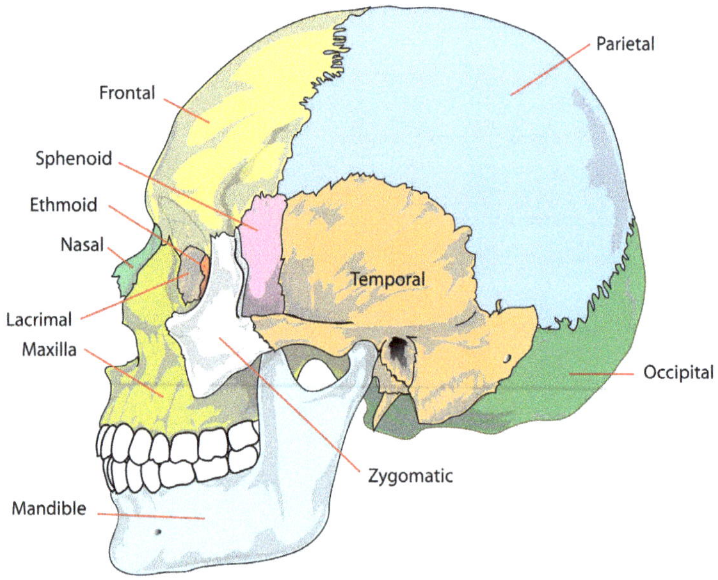

FIG. 10.7 Lateral view of the skull with major bones labelled

- **Major cavities of the skull**
  - Cranial cavity—brain
  - Orbital cavity—eyeball
  - Oral cavity—teeth and tongue

- Nasal cavity
- Paranasal sinuses
  - Frontal
  - Sphenoid
  - Ethmoidal
  - Maxillary

- **Passageways** (Things run through them, remember!)
  - Carotid canal—carotid artery
  - Cribriform foramina (CN I)—olfactory fibers
  - Foramen magnum—spinal cord
  - Foramen ovale (CN V)—mandibular branch
  - Foramen rotundum (CN V)—maxillary branch
  - Foramen spinosum—blood vessels
  - Hypoglossal canal (CN XII)
  - Inferior orbital fissure (branch of CN V)
  - Jugular foramen (CN IX, X, XI, and internal jugular vein)
  - Optic canal (CN II)—vision
  - Stylomastoid foramen (CN VII)
  - Superior orbital fissure (CN III, IV, VI, plus ophthalmic branch of CN V)

- Other essential information to know about the skull and related structures
  - Sutures—immovable joints where flat bones of the skull join
    - Sagittal suture
    - Coronal suture
    - Lambdoidal suture
    - Squamous suture
  - Sella turcica (Turkish saddle)—houses the pituitary gland
  - Cranial fossa—impressions in the contoured floor of the cranial cavity
    - Anterior—supports the frontal lobes of brain
    - Middle—supports the temporal lobes of the brain and pituitary gland
    - Posterior—supports the cerebellum and parts of the brain stem
  - Hyoid bone—inferior to the mandible
    - Only bone in the body with no direct contact with another bone
    - Frequently fractured during strangulation

NOTES

Axial Skeleton—Vertebral Column

The adult vertebral column or backbone consists of 26 bones. This includes 24 individual vertebrae and the fused vertebrae that make up the sacrum (five fused bones) and the coccyx (three, four, or five fused bones). The vertebral column supports the head and acts as an anchoring structure for the 12 ribs that form the thoracic cage.

Basement information—must-know foundational stuff about the vertebral column: Remember, we'll always be in the basement of our A&P building because we're constructing a solid foundation of A&P insight and understanding.

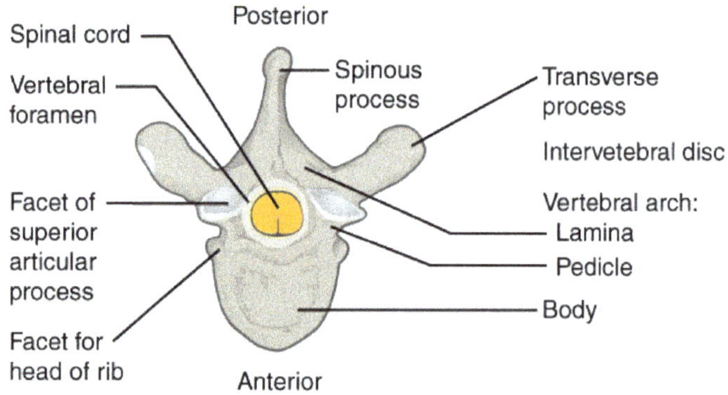

FIG. 10.8 Structure of typical thoracic vertebral body showing major components. Cervical and lumbar vertebrae will have a similar structure minus the facet for the head of the rib. C1 is only vertebrae with no body.

- **Cervical vertebrae:** Top of the spine—area of the neck
  - Total of seven: C1 through C7
  - C1 (**atlas**)—as in Greek mythology holding up the world; C1 holds up the skull, allows head to nod back and forth as in "yes"
  - C2 (**axis**)—vertical projection called dens acts as an axle; allows the head to shake side to side as in "no"
  - Transverse processes contain the transverse foramen, allowing for the passage of the vertebral arteries (left and right)
  - May have bifid (split) spinous processes (C5)
  - Smallest and lightest of all the vertebrae—structured as they are because of what they do

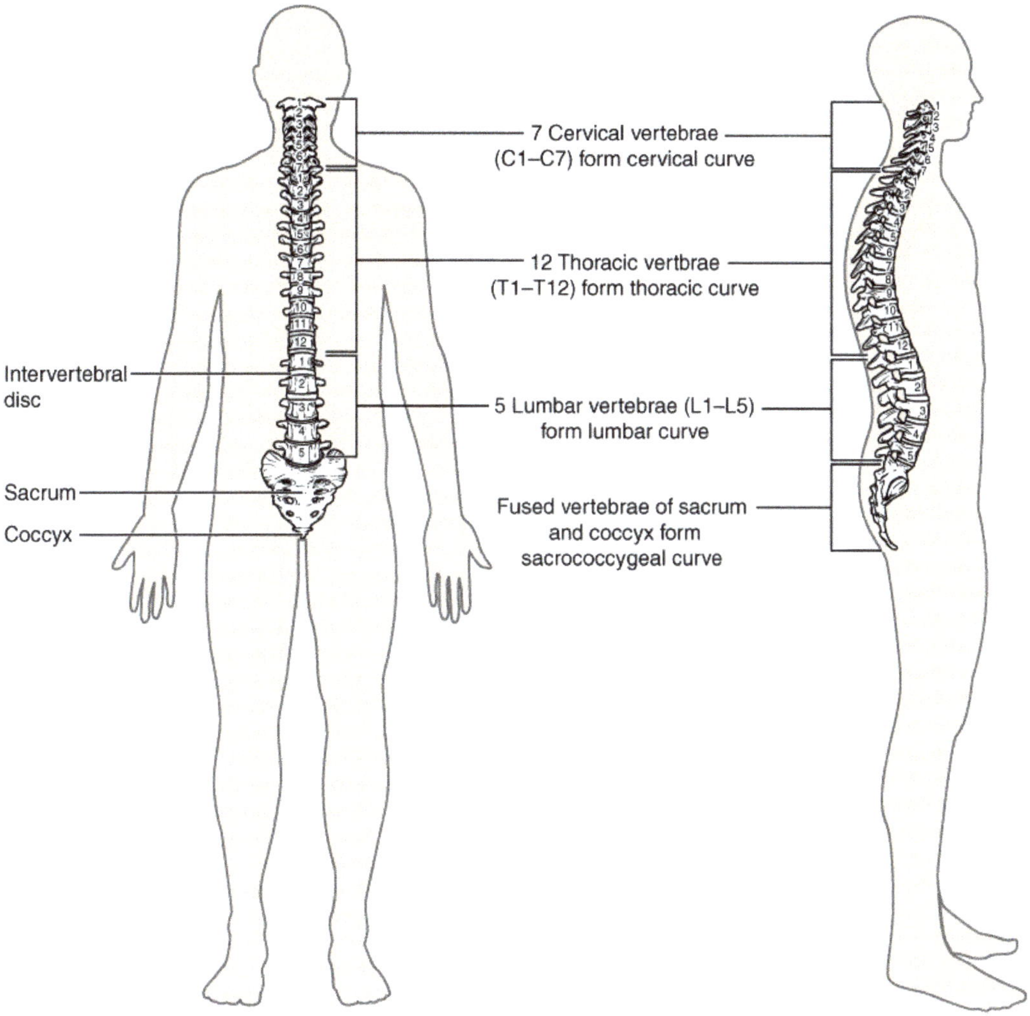

FIG. 10.9 Cervical, thoracic, lumbar, sacral, and coccyx

- **Thoracic vertebrae:** Middle of the spine—form the posterior (back) of the thoracic cage
  - Total of 12: T1 through T12
  - Have facets where the ribs join the body of thoracic vertebrae
  - Spinous processes point more inferiorly than cervical or lumbar vertebrae
- **Lumbar vertebrae:** Most inferior of the individual vertebrae
  - Large and bulky because of what they do, support everything above them
  - Support all the vertebrae and body mass superior to lumbar vertebrae
  - L5 more bulky, thicker than L1 because of what it does: supports more weight!
- **Intervertebral discs:** Located between the vertebral bodies—separate individual vertebrae
  - Cushion the spine

- Have two distinct parts
  - Annulus fibrous (fibrocartilage)—tough stuff; makes up the outer ring portion of the disc
  - Nucleus pulposus—gelatinous material
  - Discs can be compared to a jelly donut
- With age or injury, nucleus pulposus herniates through the annulus ring, causing a "slipped" or herniated disc.

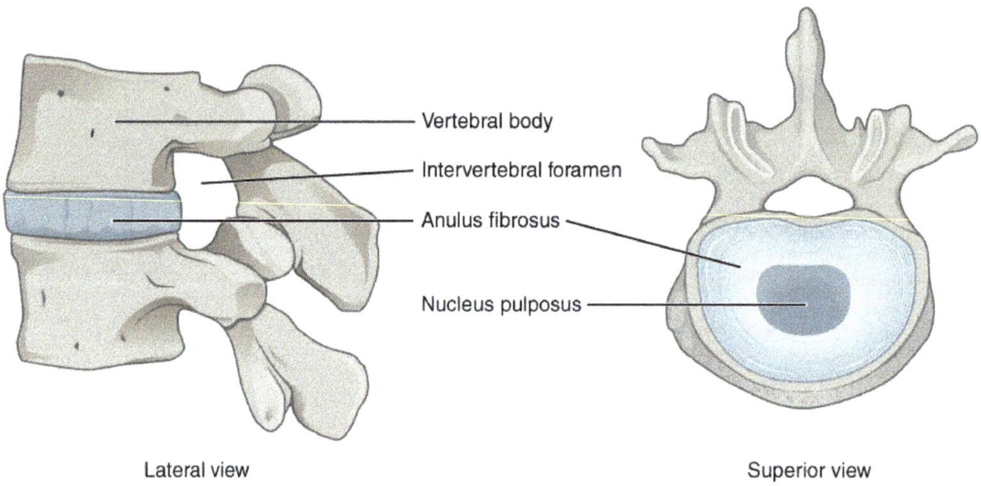

FIG. 10.10 Intervertebral disc showing annulus fibrosis and nucleus pulposus

- **Ligaments and skeletal muscles:** These run up and down the entire vertebral column to stabilize, support, and create the wide variety of spine motions
- **Sacrum:** Makes up posterior portion of the pelvis
  - Fusion of five sacral vertebrae
  - Articulates superiorly with L5 and inferiorly with the coccyx
  - Articulates laterally with the two os coxae (hip bones—left and right)
- **Coccyx:** Commonly called the "tailbone"
  - Fusion of three, four, or five coccygeal vertebrae
  - First coccygeal vertebra articulates superiorly with the sacrum
- **Vertebral curves:** Normal curves allow for equal distribution of body weight and balance in the upright position; classified as either primary or secondary
  - Primary (present shortly after birth) arch posteriorly (newborn has single "C" curve)
    - Thoracic
    - Sacral
  - Secondary (develop as child grows) arch anteriorly
    - Known as compensation curves
    - Help shift the body's weight over the lower extremities

- Cervical (develops when infant starts to raise head)
- Lumbar (develops when child starts to walk)

This is an opportunity for students to go online and search for graphics showing the three normal curvatures of the vertebral column. Remember, the ones doing the most work are the ones doing the most learning.

- **Abnormal spinal curves:** Abnormal deformities
  - **Kyphosis**—exaggerated thoracic curve
    - Directed posteriorly
    - Creates hunchback appearance (think Notre Dame)
    - Most common in postmenopausal women
    - Osteoporosis, fracture, or osteomalacia can cause kyphosis
    - Pathological in nature—caused by disease or dysfunction
  - **Scoliosis**—most common of the three abnormal spinal curvatures
    - Lateral curvature of the spine
    - Common in young females
    - Can be caused by developmental abnormalities
    - Treated with external braces or surgical intervention if severe
    - Pathological in nature—caused by disease or dysfunction
  - **Lordosis**—exaggerated lumbar curve
    - Directed anteriorly
    - Creates swayback—protrusion of abdomen
    - Abnormal but not pathological when caused by late stages of pregnancy
    - Could be caused by similar disease/dysfunction as kyphosis
    - May be caused by abdominal obesity

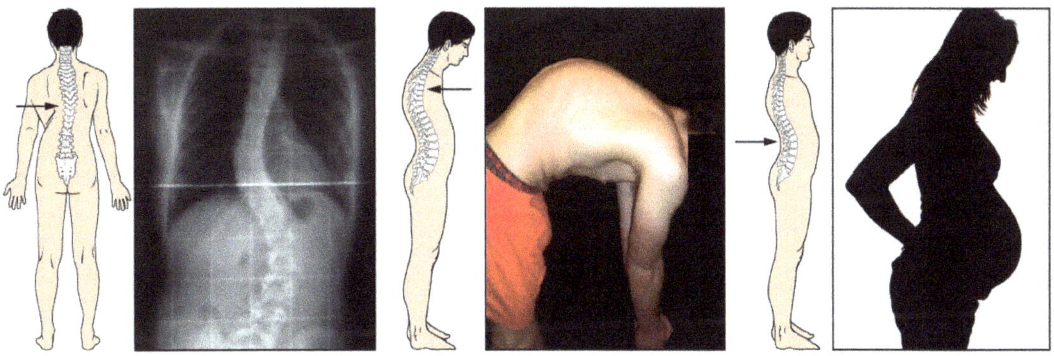

(a) Scoliosis (b) Kyphosis (c) Lordosis

FIG. 10.11 Abnormal curvatures of the spine—scoliosis, kyphosis, and lordosis

NOTES

Axial Skeleton—Thoracic Cage

This is the framework of the chest. It is made up of the thoracic vertebral column posteriorly, the ribs laterally, and the sternum (breastbone) anteriorly. The thoracic cage acts as a protective enclosure for the viscera (organs) of the chest, like the lungs and heart. It also provides attachment points for numerous muscles like the pectoralis major and minor.

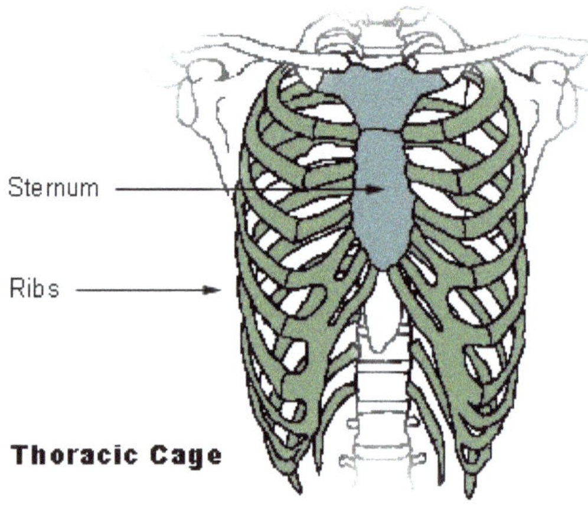

FIG. 10.12 Sternum and ribs

Foundational stuff about the thoracic cage:

- **Thoracic vertebral column**
  - T1 through T12
  - Costal facets for articulation with ribs
- **Ribs** (12 total)
  - Seven true ribs (individual ribs articulate directly with the sternum)
  - Five false ribs (cartilage of ribs 8, 9, 10 fuse with costal cartilage of rib 7)
  - Two floating ribs (ribs 11 and 12)—they don't have any connection to the sternum anteriorly
  - Head, angle and shaft
  - Costal groove on the inferior aspect of each rib contains intercostal vessels and nerves
- **Sternum** (known as the "breastbone")
  - Manubrium—superiorly; green in Figure 10.13
  - Suprasternal notch (i.e., jugular notch) at the superior aspect of the manubrium
  - Body; teal in Figure 10.13

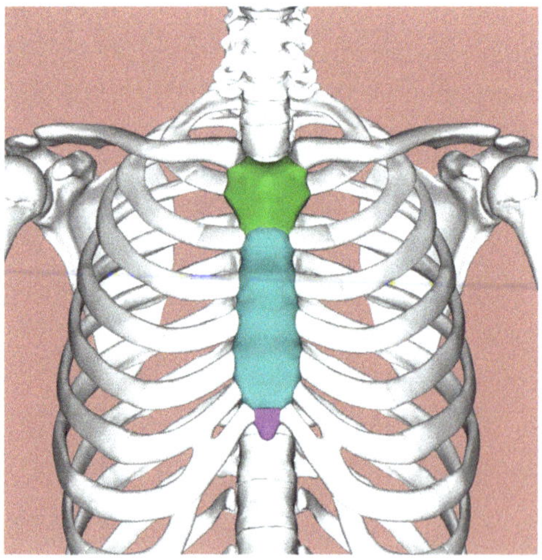

FIG. 10.13 Manubrium (green), body (teal), xiphoid process (purple)

- Sternal angle between the manubrium and body
- Xiphoid process; purple in Figure 10.13

- **Hyoid Bone**—unique, freestanding bone that does not articulate with any other bone. Only bone in the body that has this characteristic. Frequently broken in strangulation, so it has forensic pathological significance.

NOTES

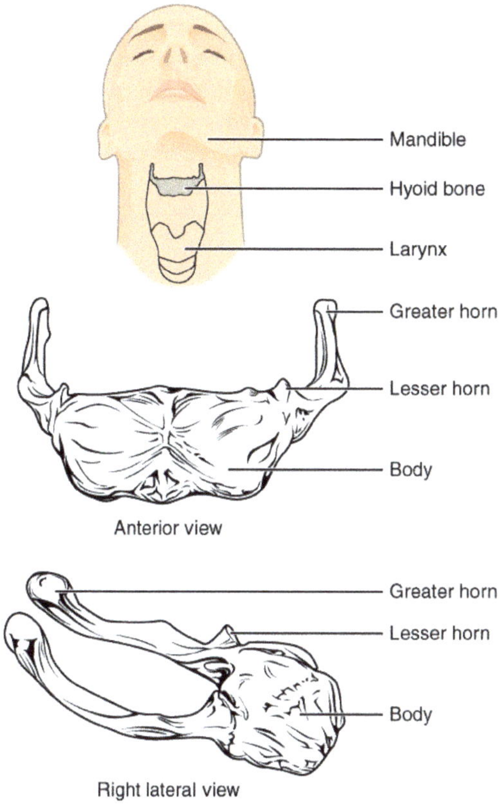

FIG. 10.14 Location, body, greater, and lesser horns

Appendicular Skeleton—Upper Extremity

The appendicular skeleton is composed of all bones that are *not* part of the axial skeleton. This includes the upper extremity, pectoral girdle (shoulder), lower extremity, and the pelvic girdle (hip).

The upper and lower extremities have some common features based on their evolutionary history. They also exhibit some differences based on their function, just another example of FFF. Things are structured the way they are because of what they do.

Both the upper and lower extremities have a "girdle" of bones that support the extremity and connect it to the axial skeleton. The pectoral girdle (scapula and clavicle) hold the upper extremity in place, and the pelvic girdle (left and right os coxae, sacrum, and symphysis pubis) secures the lower extremity.

The proximal part of each extremity contains one large bone, the humerus in the upper extremity and the femur in the lower extremity. Both extremities also contain two bones in the distal portion of the limb, the ulna and radius in the forearm and the tibia and fibula in the leg.

✔ **CONCEPT!** Utilize the similarities of the upper and lower extremities to help gain insight into their structure and function.

The upper extremity consists of the arm, forearm, wrist, and hand. The arm is the proximal part of the upper extremity. The lower extremity consists of the thigh, leg, ankle, and foot. The leg is the distal part of the limb. Anatomist and health care professionals refer to the "arm" as that part of the upper extremity from the shoulder to the elbow and the "leg" as that part of the lower extremity from the knee to the ankle. The forearm runs from the elbow to the wrist and the thigh from the hip to the knee.

Both extremities have groupings of multiple bones in the wrist (carpal) and ankle (tarsal) that allow for a variety of motions. The most distal aspects of both upper and lower extremities contain the hands and feet. Both hands and feet have similar structure, with five metacarpals (palm of the hand) and five metatarsals (arch of the foot). Finally, the hand and foot end distally with a total of 14 phalanges making up the digits of the hand (four fingers and a thumb—*not* five fingers) and the five digits or toes of the foot.

The differences between the extremities reflect their function: The lower limb is weight bearing and used for locomotion, so it sacrifices mobility (range of motion) at joints to gain stability or strength. The upper extremities are not weight bearing and have much greater range of motion, so they can perform a wide variety of tasks. Subsequently, they have much greater mobility but less stability than the lower extremity. This is another good example of FFF!

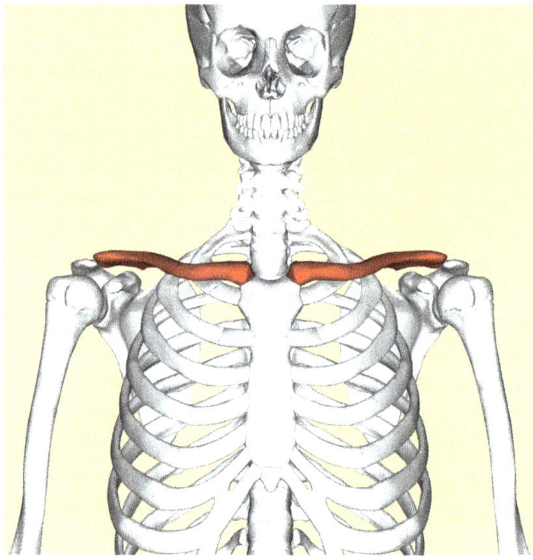

FIG. 10.15 Sternal (medial) and acromial (lateral) ends

Basement stuff—must-know foundational stuff about the appendicular skeleton: Remember we'll always be in the basement of our A&P building because we're constructing a solid foundation of A&P insight and understanding.

- **Pectoral girdle**—articulates with the trunk and supports the upper extremity
  - Only direct (boney to boney) connection between the upper extremity and the trunk in the pectoral girdle
  - Only articulation (joint) between the arm (humerus) and the trunk is the AC (acromial clavicular) joint
- **Clavicle** (Figure 10.15)—commonly known as the collar bone, it extends between the manubrium of the sternum and the acromion of the scapula
  - S-shaped with a thick sternal end and a flat acromial end
  - Easily palpated (felt) at the neck of a T-shirt
  - Subclavian groove on the inferior surface contains subclavian vessels/nerves
- **Scapula**—flat, triangular bone that is commonly known at the shoulder blade
  - Easily palpated under the skin of the upper back
  - Acromion—lateral projection joins with the clavicle forming the AC joint
  - Spine on posterior surface

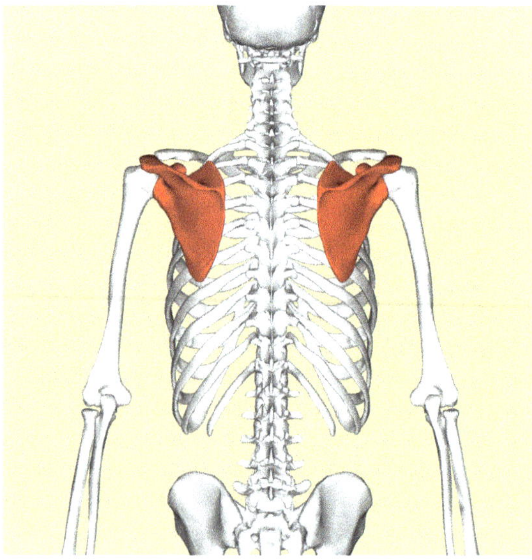

FIG. 10.16 Triangular in shape

- Supraspinous fossa (depression) superior or above the spine of scapula
- Infraspinous fossa—(depression) below or inferior to the spine of scapula
- Three borders
  - Superior—superior to spine of scapula
  - Medial—edge closest to the vertebral column
  - Lateral—most lateral and closest to the axilla (armpit)
- Glenoid cavity or fossa—shallow depression receives the head of the humerus

NOTES

- **Upper extremity**—consists of three parts
  - **Brachium**: Arm—shoulder to the elbow
  - **Antebrachium**: Forearm—elbow to wrist
  - **Hand**: Wrist to fingertips
- **Brachium**: Contains the humerus—longest and largest bone in the upper limb
  - Head of humerus—superiorly articulates with the glenoid fossa of the scapula
  - Anatomical neck of humerus—barely visible line just below the head indicating the epiphyseal plate
  - Greater and lesser tubercles—sites for attachment of skeletal muscles
    - Enlargements on the anterior surface of the humerus
    - Intertubercular sulcus (groove) formed between tubercles—contains the tendon of the long head of the biceps brachii

✔ **CONCEPT!** Where bones have enlargements such as tuberosities, ridges, lines, or trochanters, it is because something attached to the bone at that site. Those attachments stress the bone, and when bone is stressed it gets thicker, thus causing thickening and enlargement on the surface of bones.

✔ **CONCEPT!** Wolff's law states that in a healthy person, bones will respond to changes in load or stress by remodeling or reshaping. So, Wolff's law is an explanation for the presence of bumps and ridges on bones that act as sites of attachments for skeletal muscle, ligaments, or joint capsules.

- **Humerus**
  - Surgical neck—narrowing of the bone immediately distal or inferior to the tubercles
  - Called surgical neck because it is a frequent site of fracture
  - Fractures at the surgical neck often require surgical repair
- Shaft—long column of bone with deltoid tuberosity (where deltoid muscle inserts)
- Epicondyles—medial and lateral; boney projections on the distal aspect of humerus
  - Medial epicondyle—on medial side of elbow; attachment for forearm flexor muscles; flex (curl) the fingers and the hand at the wrist
  - Lateral epicondyle—on lateral side of elbow; attachment for forearm extensor muscles; extend (straighten) the fingers and the hand at the wrist
- Articular surfaces on distal humerus
  - Capitulum—located laterally; articulates with head of radius
  - Trochlea—located medially; articulates with the ulna
- Radial fossa—accommodates head of radius
- Coronoid fossa—accommodates coronoid process of ulna
- Olecranon fossa—accommodates the olecranon process of ulna

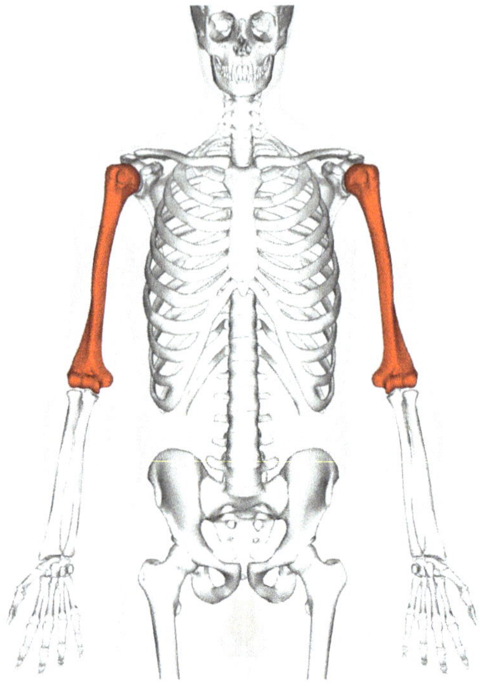

FIG. 10.17 Long Bone of the Brachium

Antebrachium—contains radius and ulna

- Radius—forearm bone on the lateral or thumb side of the forearm
  - Head—articulates with capitulum of humerus
  - Radial tuberosity—insertion point for tendon of biceps brachii
  - Shaft—long column or the body of the bone
  - Styloid process—process that can be felt on the wrist, just above the base of the thumb
- Ulna—forearm bone on the medial or little finger side of the forearm
  - Trochlear notch interlocks with the trochlea on the distal humerus
  - Coronoid process articulates with the coronoid fossa of humerus

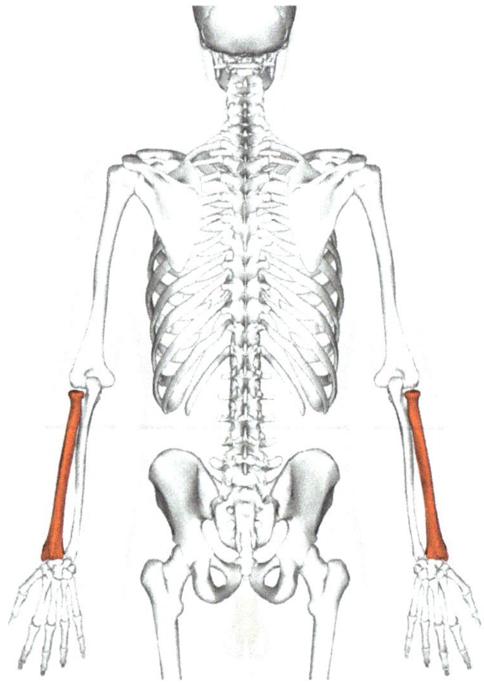

FIG. 10.18 More lateral bone of the antebrachium

- Olecranon process articulates with the olecranon fossa of the humerus
- Olecranon process makes up the point of the elbow

Hand—carpals, metacarpals, and phalanges

- Carpals—wrist bones
  - Total of eight bones
  - Short bones arranged in two rows of four bones each
  - Allows for significant variation of movement
  - At this foundational level there's no need to memorize the names of these individual bones
  - But it may be helpful if students can at least *recognize* the names of the eight carpal bones
    - Proximal row—scaphoid, lunate, triquetrum, pisiform
    - Distal row—trapezium, trapezoid, capitate, hamate

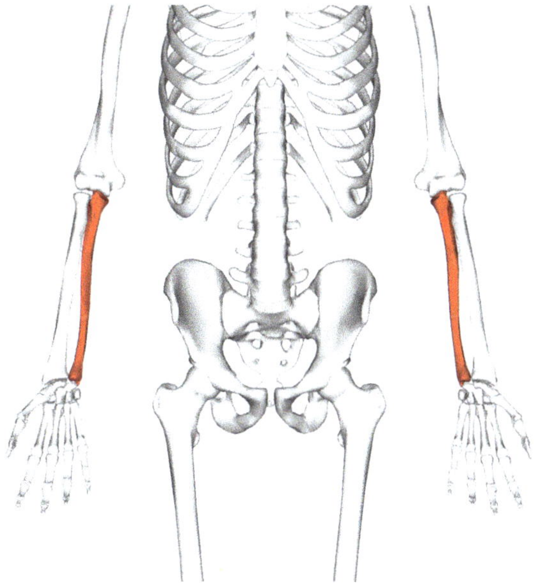

FIG. 10.19 More medial bone of the antebrachium

✔ **CONCEPT!** Understand the difference between recognizing information compared to recalling information. Filling in a blank on an answer sheet requires a student to recall from memory, whereas, picking an answer in a multiple-choice question requires an ability to recognize an item. Recall requires a greater depth of understanding than recognition.

✔ **CONCEPT!** Bones are shaped the way they are because of what they do!

- Metacarpals—make up the palm area of the hand
  - Five long, column-like bones
  - Articulate with the carpals proximally—phalanges distally
- Phalanges—comprise the digits of the hand
  - Four fingers
  - One thumb
  - ***Not five fingers!***
  - Proximal, middle, and distal phalanges
  - Two phalanges in thumb
  - Three phalanges in fingers
- MP joints—joint between metacarpals and proximal phalanges
- PIP joints—joint between proximal and middle phalanges
- DIP joints—joint between middle and distal phalanges

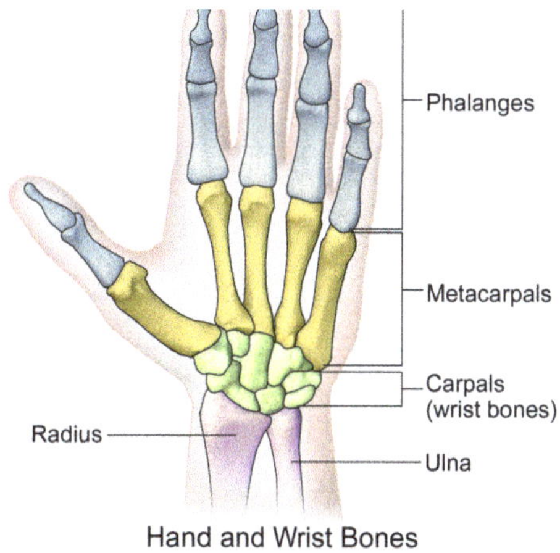

FIG. 10.20 Wrist and hand showing carpals, metacarpals, and phalanges

NOTES

Appendicular Skeleton—Pelvis

It gets a little confusing here because the pelvis contains bones of both the axial and the appendicular skeleton. The sacrum and the coccyx are part of the axial skeleton (they are part of the vertebral column) but form the posterior aspect of the pelvis. The os coxae are the hip bones, making up most of the pelvis, and are part of the appendicular skeleton. This is one of those tricky places in A&P where you've just got to bite the bullet and memorize the exceptions to general concepts.

✔ **CONCEPT!** The bones of the lower extremity are shaped the way they are because of what they do!

- **Pelvic girdle**—the adult pelvis is composed of four bones
  - Left and right os coxa
  - Sacrum
  - Coccyx
  - Symphysis pubis joins hip bones anteriorly
  - Viscera (organs or guts) in the pelvic cavity are protected by pelvis
  - Os coxa made up of three bones:
    - Ilium—most superior
    - Ischium—inferior and posterior
    - Pubic bone—inferior and anterior
  - Terms, areas of pelvic girdle to know:
    - Iliac crest—superior ridge of ilium
    - Ischial tuberosity—origin for hamstring muscles; bone that aches when you sit on a hard surface (like bleachers) for a long time
    - Acetabulum—deep pocket that receives the head of the femur
    - Pelvis brim—above this line is the **false pelvis**; below it is the **true pelvis**
    - Pelvic inlet and outlet
    - Subpubic angle

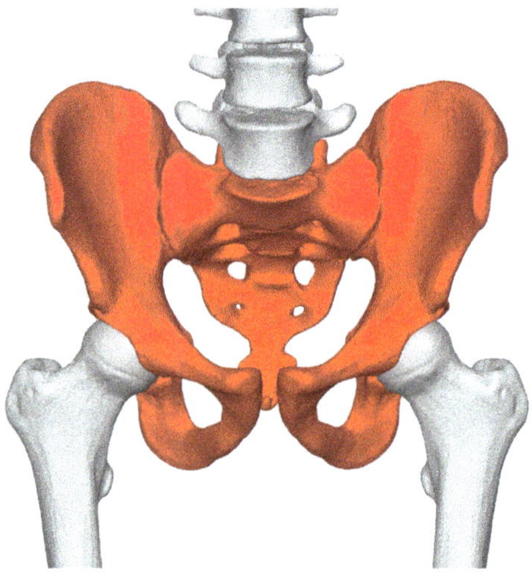

FIG. 10.21 Left and right os coxa, sacrum, coccyx, and symphysis pubis anteriorly

- Female pelvis versus male pelvis:
  - Female wider than male
  - Subpubic angle more acute in male than female
  - Pelvic inlet and outlet wider in females

✔ **CONCEPT!** The male and female pelvis are structured the way they are because of what they do. There's that FFF thing again! Females birth babies and males don't!

Appendicular Skeleton—Lower Extremity

As described earlier, the lower extremity is designed to allow an erect posture and to support the weight of the entire body. It is shaped the way it is **because of what it does.**

- **Femur** (longest and strongest bone in the body)—thigh bone
  - Head—fovea capitis (insertion site of ligamentum teres)
  - Neck
  - Greater and lesser trochanter
  - Shaft
  - Epicondyles—medial and lateral
  - Condyles—medial and lateral
  - Intercondylar fossa
- **Patella**—kneecap
  - Entirely contained in the tendon of the quadriceps muscle
  - Largest sesamoid bone in the body
- **Tibia** (largest bone in the leg)—shinbone
  - Condyles—medial and lateral
  - Forms the knee joint with the femur superiorly and the patella anteriorly
  - Tibial tuberosity (insertion point for quadriceps muscle)—remember Wolff's law?
  - Shaft
  - Medial malleolus
  - Inferior articular surface—joins with the talus bone forming the ankle joint
- **Fibula**—thin, support bone of the leg
  - Articular facet superiorly (joins with the superior, lateral part of the tibia)

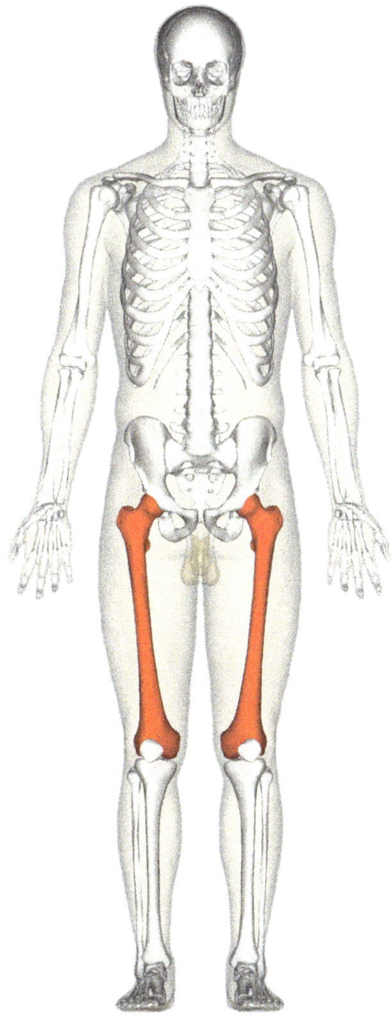

FIG. 10.22 Thigh bone—the longest, strongest bone in the body

- Head
- Neck
- Shaft
- Lateral malleolus (bump on the outside of the ankle)

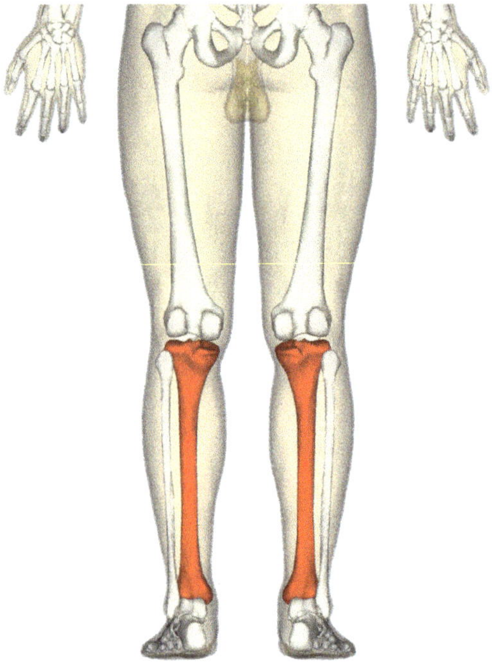

FIG. 10.23 Shin Bone—the largest bone in the leg

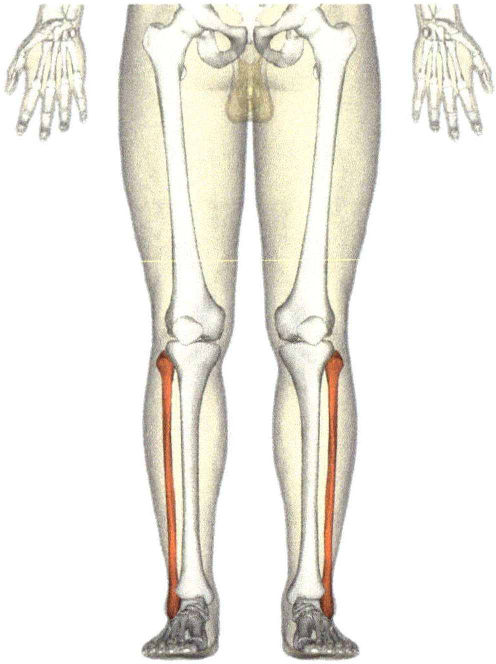

FIG. 10.24 Thin support bone of the leg. Located laterally.

- **Foot**—tarsals, metatarsals and phalanges
  - Structure is similar to hand except foot is structured to support body weight
  - Talus—articulates with distal tibia
  - Calcaneus (heel bone)—site of insertion of Achilles tendon
  - Five additional tarsal bones—need to recognize but not recall
    - Cuneiform—medial, intermediate, lateral
    - Navicular
    - Cuboid
  - Phalanges—14 as in the hand
    - Proximal, middle, and distal
    - Sesamoid bones at inferior, distal end of first phalange

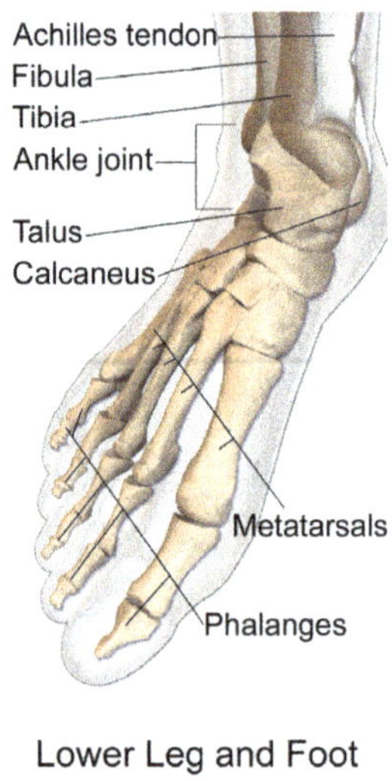

FIG. 10.25 Showing Achilles tendon, ankle joint, tarsals, metatarsals, and phalanges

NOTES

Joints—Articulations

Bones are important, but without joints between those bones there wouldn't be any movement. Articulations are the structures joining one or more bones together. These structures are frequently injured, so it is essential for anyone in a health care curriculum to gain insight into joint structure and function.

- **Joint structure**—place of contact between bones, bones and cartilage, or bones and teeth
  - Classification of joints—classified on the basis of whether a space occurs between the bones and the type of connective tissue that binds the articulating surfaces
    - **Structural classification**
      - Fibrous
        - No joint cavity
        - Held together by dense regular connective tissue
      - Cartilaginous (concept of common sense!)
        - No joint cavity
        - Bones joined by cartilage
      - Synovial
        - Fluid-filled joint cavity
        - Connective tissue capsule
        - Connected and supported with ligaments
        - Most movable joint
        - Most common joint
        - Joint of emphasis for introductory learning

      So—know your synovial joints!

      - **Functional classification**—amount of movement permitted by a joint
        - Synarthrosis—immovable joint
        - Amphiarthrosis—slightly moveable

- Diarthrosis—freely mobile joint
- Must-know joints
  - Gomphosis—between teeth and bones (mandible and maxillary bone)
  - Suture—immovable joints in the skull
  - Symphyses—pad of fibrocartilage between bones
  - Synovial—movable joints; found throughout the body
    - Shoulder
    - Knee
    - Digits—hands and feet
    - Ankles
    - Wrists
    - Frequently injured—so high clinical significance

- **Classification of synovial joints**
  - Uniaxial—moves in only one plane
  - Biaxial—moves in two planes
  - Multiaxial—moves in multiple planes

  Look at those three names; they make common sense!

✔ **CONCEPT!** Don't memorize stuff like those three classifications. **Understand them**. The names tell you exactly what they do. Gain insight and minimize memorization.

  - Six specific types of synovial joints
    - Plane joint
    - Hinge joint
    - Pivot joint
    - Condylar joint
    - Saddle joint
    - Ball-and-socket joint

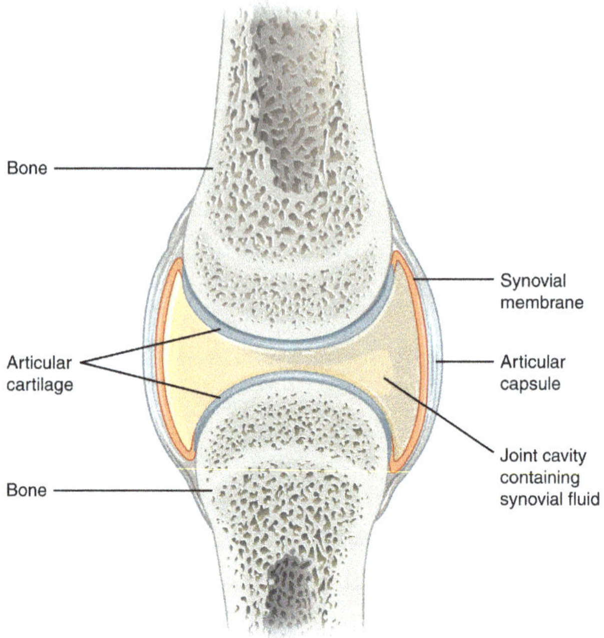

FIG. 10.26 Synovial joint showing basic structure with joint capsule

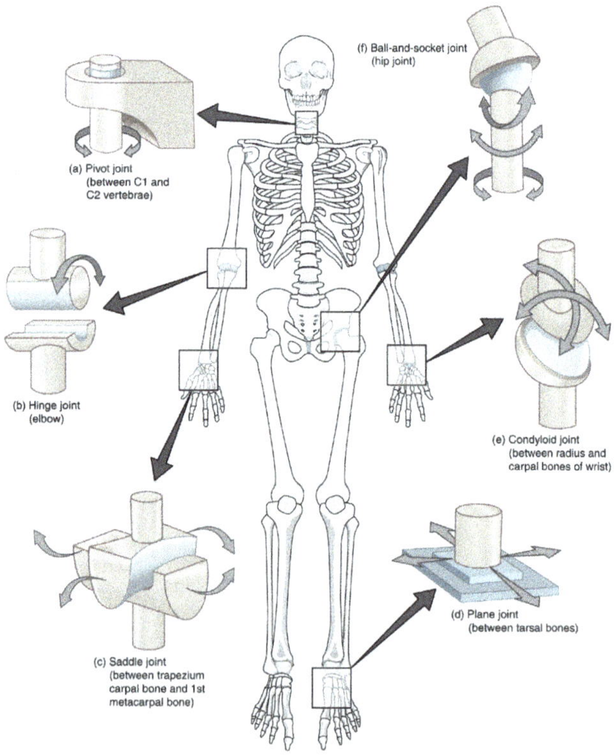

FIG. 10.27 Six types of synovial joints and their motions

- **Joints and levers**—a little physics applied to the human body
  - Three classifications of levers based on placement of fulcrum, force, and resistance
  - First-class lever—fulcrum in middle
    - Like scissors
    - Example—muscles on the posterior aspect (back) of the head and neck
  - Second-class lever—resistance in the middle
    - Like a wheelbarrow
    - Example—calf muscles on the posterior aspect of the leg
  - Third-class lever—force or effort in the middle
    - Like a forceps
    - Example—biceps brachii; flexing the forearm
    - **Most common type of lever structure in the body**

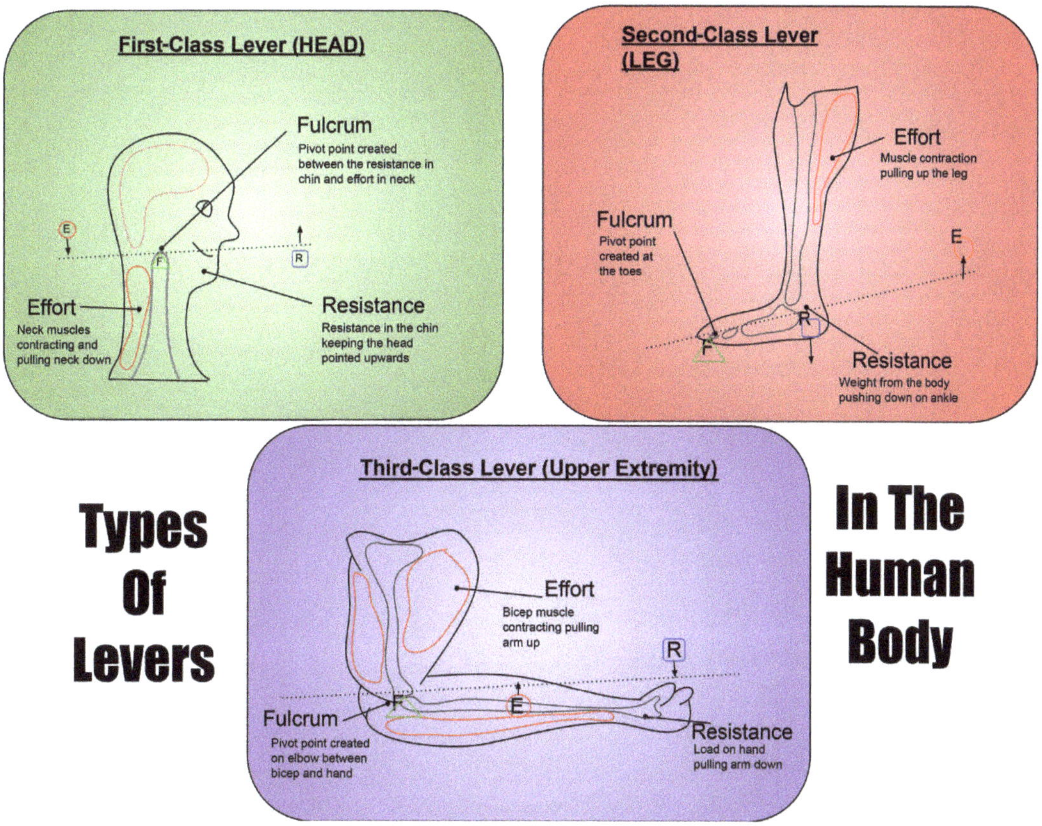

FIG. 10.28 First-, second-, and third-class levers demonstrated in the body

Here's another great place to go online and find diagrams of the three types of levers and how they apply to the human body. There are some sites that show animations of physical levers and levers in the human musculoskeletal system (Figure 10.28). **This is exactly how you CTL (churn to learn)**! Research stuff for yourself. **Struggle** and you'll learn much more than if everything is just handed to you!

✔ **CONCEPT!** The musculoskeletal systems (muscles, joints, and bones working together) provide a great opportunity for students to use their own bodies to learn. Perform the following motions as they are described or as you watch them in an online animation. Anatomy and physiology are all about you!

- **Joint motions**—must-know stuff about motions (use your own body to demonstrate motions):
  - Essential for you to look these up in a reference text or online (research them); find pictures and/or animations and get comfortable with them
  - Flexion, extension, hyperextension of the head
  - Flexion and extension
    - Forearm at the elbow
    - Leg at the knee (Remember anatomically/medically the leg is that part of the lower extremity from the knee to the ankle)
    - Circumduction of both extremities
    - Abduction and adduction—away from or toward the midline
      - Upper extremity
      - Lower extremity
      - Digits of the hand
    - Rotation—medially (toward the midline) and laterally (away from the midline)
      - Head
      - Upper extremity
      - Lower extremity
    - Pronation and supination of the hand
      - Supination—palm upward like holding a soup bowl
      - Pronation—palm downward like emptying (pouring) out a soup bowl
    - Dorsiflexion and plantar flexion of the foot
    - Inversion and eversion of the foot
    - Opposition (thumb and fingers)—unique to humans

- **Elbow joint**—know the **annular ligament**
  - Holds the head of the radius in place

- Frequent site of dislocation of the head of the radius in young children when pulled upward by their hands by a loving family member just having fun but can appear abusive

- **Hip joint**—secure joint that sacrifices range of motion for stability

✔ **CONCEPT!** Many ligaments are named by the bones they connect. Reduce memorization by taking advantage of the fact A&P are logical sciences. Terms make sense and tell you about the structure. This is particularly true in the skeletal and muscular systems. Use this to your advantage. **Understand** and there is less need to memorize.

- Ligamentum teres—ligament of head of femur
  - Teres inserts into the fovea capitus
  - Intracapsular (inside the joint capsule) ligament which is rare
  - Teres runs between the head of the femur and the acetabulum of pelvis
  - Extra stability provided to hip joint

- **Knee Joint**—this is the joint of emphasis for the novice A&P student
  - Classic synovial, diarthrotic joint
  - Frequently injured in sports and daily activities
  - Need to know **13 structures** of the knee joint
    - Femur
    - Tibia
    - Fibula
    - Quadriceps tendon
    - Patella
    - Patellar ligament
    - Tibial tuberosity
    - Medial collateral ligament
    - Lateral collateral ligament
    - Anterior cruciate ligament
    - Posterior cruciate ligament
    - Medial meniscus
    - Lateral meniscus

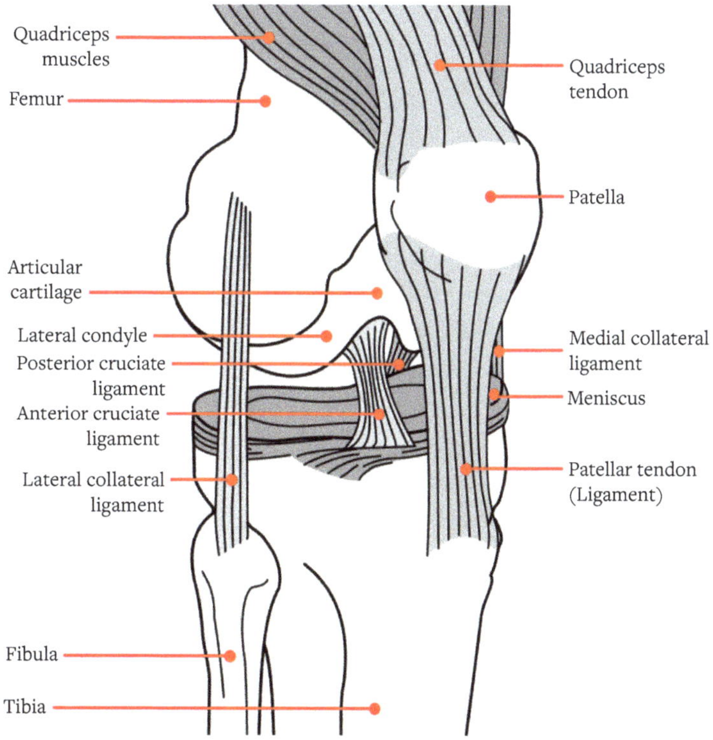

FIG. 10.29 Knee joint graphic showing some of the thirteen need to know structures.

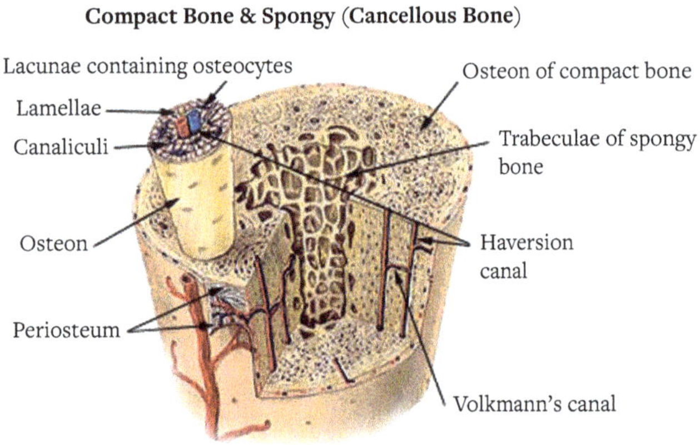

FIG. 10.30 Cross-section of long bone showing structures and osteon

Bone Physiology and the Homeostasis of Calcium

The human skeleton provides structure, support, and protection for the body. It is the major storage site for calcium, which is essential not only for bone strength but for normal function of all three types of muscle (cardiac, visceral, and skeletal), nerve function (synapses), and blood clotting.

The structural and functional unit of compact bone is the **osteon**. Histologically, the osteons in compact bone look like the stump of a tree that has been sawed down, showing its growth rings (see Figure 10.30).

BRIDGE to Chapter 8, Histology, Fig. 8.2 on page 78 to see slide of osseous compact bone

Osteocytes (bone cells) are contained in the lacunae of the osteon. Osteocytes are mature bone cells that maintain the integrity of compact bone. They evolved from osteoblast cells. Compact bone also contains two types of bone cells important for the A&P novice to know:

- OsteoBlasts—derived from primitive bone cells
  - Build bone (that's why the "B" is capitalized)—**blasts build bone**
  - Take calcium out of the blood stream and deposit it into the bone matrix
- OsteoClasts—derived from monocyte like blood cells
  - Chew bone (that's why the "C" is capitalized)—**clasts chew bone**
  - Resorbs calcium from the bone matrix and transfers it into the blood stream

BRIDGE to Chapter 3 page 25 for a discussion about homeostasis

It is imperative for the student to gain insight into the importance of calcium in human physiology. This means an understanding of how the body maintains the homeostatic balance of blood calcium. This is one of the most important TOEs in this guide. The homeostatic balance of both blood glucose (primary energy source) and blood calcium need to be an ongoing discussion in an introductory A&P course. So, let's start with a more detailed look at homeostasis, or the big "H" word. Homeostasis was mentioned back in chapter 3 but now requires a more detailed discussion.

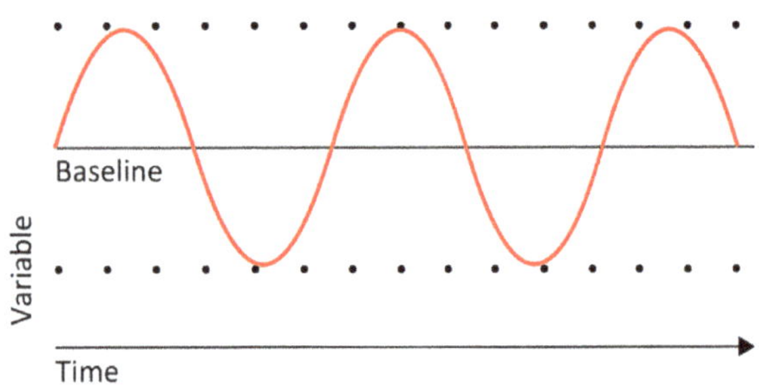

FIG. 10.31 Demonstrating dynamic constancy of calcium

One effective way to define homeostasis is **dynamic constancy** or **DC**. This describes a state of continual fluctuation, but the changes stay within an upper and a lower limit, constantly moving around a set point. It's helpful if we look at this as a graph showing a sine wave, a mathematical curve describing a smooth repetitive oscillation. No, you don't need to worry about this turning into a math text; it's just helpful to be able to visualize (REVV CCCAMM) homeostasis. In the body, there is constant fluctuation, but it's not necessarily a smooth oscillation. Depending on dozens of variables, the fluctuation may vary dramatically. When values oscillate outside the upper and lower limits the body experiences dysfunction or disease.

This type of fluctuation within an upper and lower limit applies to many bodily functions, like temperature, blood calcium, blood glucose, acidity, and many others.

NOTES

Our bodies maintain dynamic constancy by utilizing negative feedback loops. It's easy to understand these loops by relating them to something all of us know. Here's REVV CCCAMM again, with the "A" referring to the association of new material to existing knowledge. The HVAC (heating, ventilation, air conditioning) system in buildings and homes is a good example of negative feedback loops. If you graphed the temperature swings in a home, you'd see a sine wave showing dynamic constancy (DC).

The thermostat is set to a desired temperature. This is the set point or baseline on the sine wave graph. In the warm months when the temperature in the home rises above the set point, the system turns on the AC, cooling the home. Once the temperature goes slightly past the set point, the AC turns off until the temperature exceeds the set point again. This creates the oscillation of DC. In the cold months the HVAC would turn the heat on and off to maintain a dynamic constancy. **Negative feedback systems always reverse the original stimulus** (temperature change).

✔ **CONCEPT!** **Negative feedback** always **reverses** the original stimulus.

The body also utilizes positive feedback loops, which you might guess amplify the original stimulus. (More about these positive loops when we get to the endocrine system.)

The body uses an antagonistic pair of endocrine hormones to maintain the DC of blood calcium, which is essential for bone growth, muscle, and nervous function, and blood clotting.

Parathyroid hormone, produced by the parathyroid glands (makes perfect sense), raises blood calcium levels. Calcitonin produced by the thyroid gland lowers blood calcium levels. Conceptually, this process is summarized as follows:

- Low blood calcium levels
  - Chief cells in the parathyroid gland sense low blood calcium levels
  - Chief cells produce parathyroid hormone (PTH)
  - PTH is secreted by **exocytosis** (remember that?) into the interstitial space
  - PTH accumulates and creates a concentration gradient
  - PTH diffuses down the concentration gradient and into the blood stream

- PTH targets three effectors
  - Bone—stimulating osteoclasts to resorb bone and release calcium
  - Kidney—decreasing excretion of calcium in the urine
  - Small intestine—stimulating vitamin D, which in turn causes increased absorption of calcium from the gut

- High blood calcium levels
  - Parafollicular cells in the thyroid gland sense high blood calcium levels
  - Calcitonin is released into the interstitial space
  - Concentration gradient of calcitonin builds in interstitial space
  - Calcitonin diffuses into blood stream and targets the same three effectors as PTH
    - Bone—stimulates osteoblasts to build calcium into the bone matrix
    - Kidney—increasing excretion of calcium in the urine
    - Small intestine—decreasing absorption of calcium from the gut

Although PTH and calcitonin act as an antagonistic hormone pair, PTH is the more dominate chemical and the primary source maintaining calcium homeostasis.

Phosphate also plays a key role in maintaining calcium balance and bone strength. Calcium and phosphate levels are related to each other, but to simplify we've focused only on calcium homeostasis.

NOTES

Review Questions

- What does the human skeleton do?
- How does FFF relate to the skeleton?
- Define the axial and appendicular skeleton.
- What is an articulation?
- What are the six specific motions at joints?
- What are the two ways to classify joints?
- What are the conceptual functions of the skeleton?
- What is the functional unit of compact bone?
- What is a ball-and-socket joint?

- What are abduction and adduction of the upper extremity?
- What is opposition?
- Calcium is important in what four specific areas of human physiology?
- What do osteoblasts do to bone?
- What do osteoclasts do to bone?
- What endocrine hormone raises blood calcium levels?
- How does the body maintain the homeostatic balance of calcium?

Activities

- Get out those colored pencils and/or crayons again. Find a good image of the human knee online, in *SS A&P*, or in an A&P coloring book. Or, even better, draw (freehand) out an image of the knee. Color in the 13 different structures you need to know about the knee. Work on this until you can do it comfortably and with confidence from memory!
- Draw a diagram demonstrating DC, or dynamic constancy. This could be a sine wave, but be creative and come up with another way to picture DC. Maybe a teeter-totter?
- Draw a map showing the negative feedback loop system that lowers blood calcium levels utilizing PTH. You can utilize a traditional reference text or the internet to help with this activity.
- Explain calcium homeostasis to someone who doesn't understand it.

FIGURE CREDITS

Figure 10.1: Copyright © by OpenStax College (CC BY 3.0) at https://commons.wikimedia.org/wiki/File:601_Bone_Classification.jpg.

Figure 10.2: Copyright © GnolizX (CC by 3.0) at https://commons.wikimedia.org/wiki/File:603_Anatomy_of_Long_Bone_zh.jpg.

Figure 10.3: Copyright © by OpenStax College (CC BY 3.0) at https://commons.wikimedia.org/wiki/File:607_Periosteum_and_Endosteum.jpg.

Figure 10.4: Source: https://commons.wikimedia.org/wiki/File:Axial_skeleton_diagram.svg.

Figure 10.5: Source: https://commons.wikimedia.org/wiki/File:Appendicular_skeleton_diagram.svg.

Figure 10.6: Copyright © by CC BY-SA 3.0 () at https://upload.wikimedia.org/wikipedia/commons/thumb/5/5b/Skull_foramina_labeled.svg/1280px-Skull_foramina_labeled.svg.png.

Figure 10.7: Source: https://commons.wikimedia.org/wiki/File:Human_skull_side_simplified_(bones).svg.

Figure 10.8: Copyright © by BodyParts3D/Anatomography (CC BY-SA 3.0) at https://commons.wikimedia.org/wiki/File:Vertebra_Superior_View.jpg.

Figure 10.9: Copyright © OpenStax College (CC BY 3.0) at https://commons.wikimedia.org/wiki/File:715_Vertebral_Column.jpg.

Figure 10.10: Copyright © by OpenStax College (CC BY 3.0) at https://commons.wikimedia.org/wiki/File:716_Intervertebral_Disk.jpg.

Figure 10.11: Copyright © by OpenStax College (CC BY 3.0) at https://commons.wikimedia.org/wiki/File:717_Abnormal_Curves_of_Vertebral_Column.jpg.

Figure 10.12: Source: https://commons.wikimedia.org/wiki/File:Illu_thoracic_cage.jpg.

Figure 10.13: Copyright © by Anatomography (CC BY-SA 2.1 JP) at https://commons.wikimedia.org/wiki/File:Sternum_composition.png.

Figure 10.14: Copyright © OpenStax College (CC BY 3.0) at https://commons.wikimedia.org/wiki/Category:Hyoid_bones#/media/File:712_Hyoid_Bone.jpg.

Figure 10.15: Copyright © Anatomography (CC BY-SA 2.1 JP) at https://commons.wikimedia.org/wiki/File:Clavicle_-_anterior_view.png.

Figure 10.16: Copyright © Anatomography (CC BY-SA 2.1 JP) at https://commons.wikimedia.org/wiki/File:Scapula_-_posterior_view.png.
Figure 10.17: Copyright © Anatomography (CC BY-SA 2.1 JP) at https://commons.wikimedia.org/wiki/File:Humerus_-_anterior_view.png.
Figure 10.18: Copyright © Anatomography (CC BY-SA 2.1 JP) at https://commons.wikimedia.org/wiki/File:Radius_-_posterior_view.png.
Figure 10.19: Copyright © Anatomography (CC BY-SA 2.1 JP) at https://commons.wikimedia.org/wiki/File:Ulna_-_anterior_view2.png.
Figure 10.20: Copyright © BruceBlaus (CC BY 3.0) at https://commons.wikimedia.org/wiki/File:Blausen_0440_HandBones.png.
Figure 10.21: Copyright © DBCLS (CC BY-SA 2.1 JP) at https://commons.wikimedia.org/wiki/File:Pelvis_(male)_02_-_anterior_view.png.
Figure 10.22: Copyright © Anatomography (CC BY-SA 2.1 JP) at https://commons.wikimedia.org/wiki/File:Femur_-_anterior_view.png.
Figure 10.23: Copyright © Anatomography (CC BY-SA 2.1 JP) at https://commons.wikimedia.org/wiki/File:Tibia_-_posterior_view.png.
Figure 10.24: Copyright © Anatomography (CC BY-SA 2.1 JP) at https://commons.wikimedia.org/wiki/File:Fibula_-_anterior_view.png.
Figure 10.26: Copyright © OpenStax College (CC BY 3.0) at https://commons.wikimedia.org/wiki/File:907_Synovial_Joints.jpg.
Figure 10.27: Copyright © by OpenStax College (CC BY 3.0) at https://commons.wikimedia.org/wiki/File:909_Types_of_Synovial_Joints.jpg.
Figure 10.28: Copyright © by JG6115 (CC BY-SA 4.0) at https://commons.wikimedia.org/wiki/File:Levers_in_the_Human_Body.pdf.
Figure 10.29: Source: https://commons.wikimedia.org/wiki/File:Knee_diagram.png.
Figure 10.30: Source: https://commons.wikimedia.org/wiki/File:Illu_compact_spongy_bone.jpg.
Figure 10.31: Source: Mcgraw-Hill Education.

Churn to Learn!

Teach to Learn!

CHAPTER ELEVEN

The Muscular System and Muscle Contraction Physiology

TOES (TOPICS OF EMPHASIS)

- Types of muscle
- Organizational patterns of skeletal muscle
- Muscle nomenclature and structure
- Synergistic and antagonistic muscles
- Unique muscles of the body
- Motor units
- Neuromuscular junctions
- Actin (thin filament) and myosin (thick filament)
- Muscle actions around joints
- Sliding filament "theory" of muscle contraction

Look at the H_2O sheet in the systems column where it states muscle tissue creates movement of most everything in the body! When things move in the body (with a rare exception), muscle tissue is involved.

Skeletal Muscles

The body contains three different types of muscle: cardiac muscle, found only in the heart; visceral, smooth or involuntary muscle (you need to know all three names!), found in the viscera (guts) such as the walls of hollow organs like the intestine, airways of the lungs, and arteries; and skeletal, striated, or voluntary muscle (again, you need to know all three names!), which we'll focus on in this chapter.

Skeletal muscles attach to the bones of the skeleton, or other structures, and cause movement. Our skeletal muscles move the body, eyes, and vocal cords, control urine and feces, and provide our ability to engage in sexual activity. Skeletal muscle is also called striated because, histologically, it appears striped, as we learned in chapter 8. Skeletal muscle is under our conscious, voluntary control.

BRIDGE to Chapter 8, Histology, Fig. 8.3 on page 78 for slides of muscle tissue

✔ **CONCEPT!** You can minimize memorization by understanding that if you can voluntarily control a muscle it must be composed of skeletal, striated, or voluntary muscle!

✔ **CONCEPT!** The only active thing muscle can do is contract!

Skeletal muscle, in combination with the skeleton, forms what is called the musculoskeletal system. Bones and muscles depend on each other to function. Our H_2O sheet indicates this complementary nature in the fourth column with a white arrow running between muscular and skeletal systems.

Skeletal muscles are wrapped into bundles of three groupings of muscle, cells, each surrounded by a connective tissue sheet.

- Endomysium
    - Surrounds each individual muscle fiber or cell
    - Thin and delicate membrane
    - Loose areolar connective tissue
- Perimysium
    - Surrounds groups of muscle cells called fascicles
    - Dense irregular connective tissue
    - Rich in blood vessels and nerves
- Epimysium
    - Dense regular connective tissue
    - Surrounds the entire muscle

Organizational Structure of Skeletal Muscle

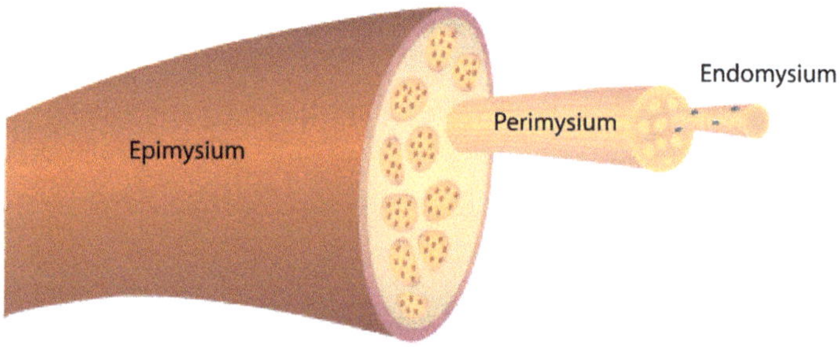

FIG. 11.1 Graphic demonstrating endomysium, perimysium, and epimysium

Tendons are formed as the three layers of wrappings merge and then extend past the muscle body. Tendons cross joints and connect skeletal muscle to bones by inserting into the periosteum of bone, providing movement at those joints.

Organizational Patterns of Skeletal Muscle Fibers

Skeletal muscle fibers are made up of bundles of muscle fibers, called fascicles, which lie parallel to each other within each muscle. However, depending on the specific function of a muscle, the organization of fascicles may vary.

Remember, a structure, including a muscle, is shaped the way it is because of what it does.

Students need to be aware there are four different patterns of fascicle arrangement:

- Circular—orbicularis oris
- Parallel—abdominal rectus
- Convergent—pectoralis major

- Pennate—which has three subdivisions:
  - Unipennate—extensor digitorum
  - Bipennate—rectus femoris
  - Multipennate—deltoid

Do an online search to find images of these different muscle arrangement patterns.

✔ **CONCEPT!** Because the only active thing muscle can do is contract or shorten, see if you can visualize what happens when each of the fascicle arrangements contract. Think about it! Talk with a study buddy about it! Draw the muscle out and then contract it to see how your drawing changes. Have some fun with it! Remember, emotions (good or bad) help you learn.

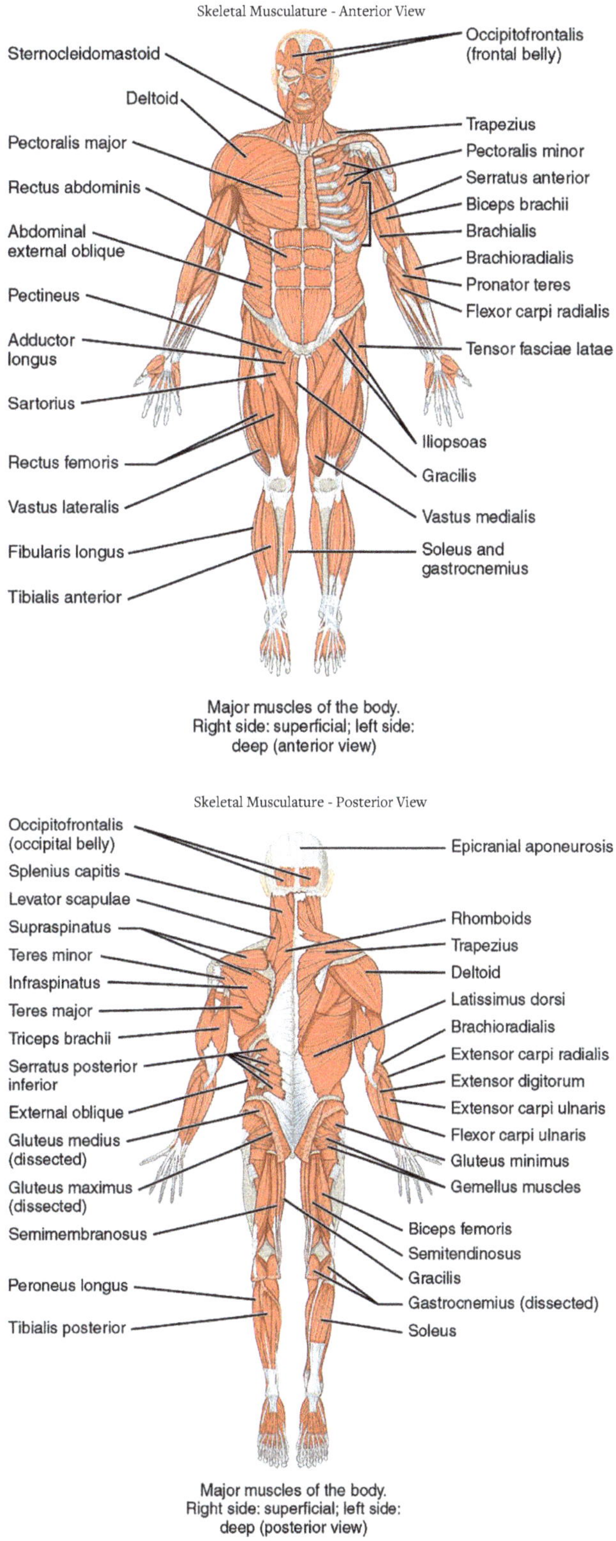

FIG. 11.2 Anterior and posterior views

Naming of Skeletal Muscles

At an introductory level, it is important for students to understand the seven different ways skeletal muscles are named. With this insight, students can then deduce the likely patterns of muscle arrangements, actions, origins, and insertions. This will help to minimize the need to mindlessly memorize isolated facts regarding skeletal muscles.

As stated, skeletal muscles are named in one of seven different ways. It is essential for the introductory student to gain insight into these means of naming muscles. The names will reveal specific facts about muscles, such as origins, insertions, actions, size, shape, and location.

Muscle Naming

- **Muscle action**—what the muscle does!
  - Flexor—flexor carpi radialis
  - Extensor—extensor hallucis longus
  - Abductor—abductor pollicis longus
  - Adductor—adductor magnus

- **Muscle attachment**—origin (beginning) and insertion (ending)
  - Origin is listed first in the name—less moveable part of skeleton
  - Insertion is listed second in the name—more moveable part of skeleton
  - Sternocleidomastoid—sternum, clavicle, and mastoid process of skull
- **Specific body regions**—This is why we learned anatomical regions in Chapter 4 of *SS A&P*
  - Brachii—arm
  - Femoris—thigh
  - Hallicus—great toe
  - Pollicus—thumb
- **Shape of muscle**—how the muscle appears
  - Deltoid—triangular
  - Trapezius—trapezoidal
  - Longus—long
  - Brevis—short
- **Orientation of muscle fibers**—organizational patterns
  - Rectus (straight)—rectus abdominus
  - Oblique (angled)—external or internal oblique
- **Size of muscle**—big or small
  - Major—pectoralis major
  - Minor—pectoralis minor
  - Maximus—gluteus maximus
  - Medius—gluteus medius
  - Minimus—gluteus minimus
- **Muscle heads**—number of muscle heads at the origin of the muscle
  - Biceps—two heads; biceps brachii
  - Triceps—three heads; triceps brachii
  - Quadriceps—four heads; quadriceps femoris

NOTES

Types of Skeletal Muscles

Muscles usually work collaboratively to produce movements. Generally, they do not work in isolation. There are three basic muscle types students need to know. These are based on the primary function of the muscle.

- **Agonist**—prime mover
  - Muscle contraction causes specific movement
  - Biceps brachii of the upper extremity produces flexion at the elbow
- **Antagonist**—muscle whose movement opposes the action of an agonist
  - Agonist produces flexion, then antagonist produces extension
  - Biceps brachii produces flexion of the forearm at the elbow
  - Triceps brachii produces extension of the forearm at the elbow
  - Antagonistic muscle pairs move body parts in opposite directions
- **Synergist**—muscle that assists an agonist's action
  - Biceps brachii and brachialis—work together to cause flexion at the elbow
  - May act to stabilize the site of origin of another muscle—called a **fixator**
  - Serratus anterior fixates the scapula during a punching type of motion

There are an overwhelming number of skeletal muscles, and most introductory A&P students are required to memorize many of those muscles and specific details about those muscles.

But remember, this guide is focused on the basement of the four-story building we're constructing. All those details about skeletal muscles are first- and second-floor facts!

There are at least fifty-six muscles and connective tissue structures in Figure 11.2. But at an introductory level the student needs to focus on thirty to thirty-five of the major and most superficial skeletal muscles as listed on pages 55 and 56 of the *SS A&P* Lab Manual, 1st Edition. It is also helpful to examine skeletal muscle groups.

Skeletal Muscle Groups

These are synergistic and antagonistic muscle groups or pairs the novice A&P student needs to learn.

Synergistic muscle groups—collaborate to produce motions

- **Muscles of mastication**—muscles of the head involved in chewing
  - Temporalis
  - Masseter
  - Medial pterygoid
  - Lateral pterygoid
  - Buccinators—some anatomists don't include this muscle in the chewing group, but we will in this guide. (*Yes*, there are differences of opinion in A&P!)
  - Buccinator muscles can also be considered one of the muscles of facial expression.

- **Rotator cuff group**—involved in a throwing motion (pitcher in baseball)
  - Supraspinatus
  - Infraspinatus
  - Subscapularis
  - Teres minor
- **Abdominal muscles**—form the anterior wall of the trunk, from superficial to deep
  - External oblique
  - Rectus abdominus
  - Internal oblique
  - Transverse abdominus
- **Flexors of the forearm at the elbow**
  - Brachialis
  - Biceps brachii
  - Brachioradialis
- **Thigh extensor/abductor group—the "glutes"**
  - Gluteus maximus
  - Gluteus medius
  - Gluteus minimus
- **Thigh adduction group**
  - Adductor longus
  - Adductor magnus
  - Adductor brevis
  - Gracilis
  - Pectineus
- **Hamstrings group**
  - Biceps femoris
  - Semitendinosus
  - Semimembranosus
- **Calf muscles**—plantar flex or point the toes downward (standing on tiptoes)
  - Gastrocnemius
  - Soleus

Antagonist muscle pairs—oppose the action of the other muscle

- **Biceps and triceps brachii**
  - Biceps, brachialis, and brachioradialis flex forearm at the elbow
  - Triceps extend forearm at the elbow

- **Pectoralis major and latissimus dorsi**
  - Pecs—principle flexor of the arm at the shoulder
  - Lats—principle extensor of the arm at the shoulder
- **Quadraceps femoris and hamstrings**
  - Quads—extend the leg at the knee
  - Hamstrings—flex the leg at the knee

✔ **CONCEPT!** Minimize memorization by recognizing muscle groups and compartments. As the synergistic groups and antagonist pairs demonstrate, muscles are bunched together to create effective movement of body parts. Focus first on these groupings to build a solid foundation of understanding. Then it will be easier to add details as you progress through more advanced courses and into the health care workforce.

Muscle Compartments

- **Forearm flexors**
  - Primarily flex the hand at the wrist and flex the fingers
  - Many originate from the medial epicondyle of the humerus
- **Forearm extensors**
  - Primarily extend the hand at the wrist and extend the fingers
  - Many originate from the lateral epicondyle of the humerus
- **Compartments of the thigh**
  - Medial compartment—adducts the thigh (pulls thigh toward the midline), pulls thighs together
  - Anterior compartment (quads)—extends (straightens) the leg at the knee
  - Posterior compartment (hamstrings)—flexes (bends) the leg at the knee

NOTES

External Ocular Muscles

There are six skeletal muscles surrounding the outside of the eyeball, which allow movement of the eye. You can easily deduce these extraocular muscles are skeletal muscle because they are under our voluntary control. They are voluntary muscle, another name for skeletal or striated muscle.

✔ **CONCEPT!** If you can voluntarily contract a muscle it is composed of voluntary or skeletal muscle.

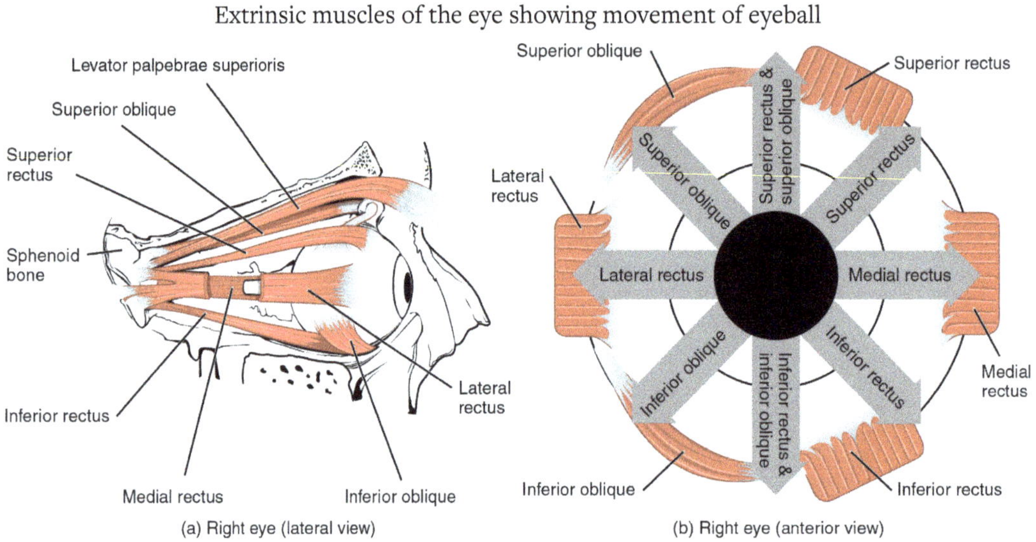

FIG. 11.3 Voluntary/skeletal muscles

Novice A&P students need to know all six of the striated extraocular muscles surrounding the eyeball, which allow us to voluntarily move our eyes.

- Four rectus (straight) muscles
  - Superior rectus
  - Medial rectus
  - Inferior rectus
  - Lateral rectus
- Two oblique (angled) muscles
  - Superior oblique
  - Inferior oblique

We'll go into more detail about the eye in chapter 12. The eye is mentioned here because the muscles on the outside of the eye are voluntary or striated muscle. There are also three muscles inside the eye. These intrinsic eye muscles are visceral (smooth) or involuntary muscles.

Unique and unusual muscles are muscles that have unique characteristics or names based on their function or the fact they typically become over developed (larger) in specific professions.

- **Boxers muscle**
  - Serratus anterior
  - Stabilizes the scapula during a punching motion
  - Overdeveloped in someone who punches for a living, like a boxer
- **Tailor's muscle**
  - Sartorius
  - Longest muscle in the body
  - Old-time tailors who sat crossed-legged while sewing overdeveloped their sartorius
- **Fright muscle**
  - Platysma muscle
  - Only muscle in the body with minimal to no boney attachments
  - Primarily runs from skin to skin
  - Pulls mandible inferior when making a frightened face

✔ **CONCEPT!** There are so many muscles and so much to learn about the muscular system that it is imperative at an introductory level to minimize the amount of content and focus on concepts and the major muscle groups. Absorbing additional details will come during future courses and with experience. This guide reduces the amount of content so students can do more repetition to facilitate learning and retention. Remember, REVV CCCAMM!

NOTES

Muscle Physiology and the Sliding Filament "Theory" of Contraction

Types of Muscle

As the H_2O sheet shows, muscle is one of the four primary tissues. **Remember E, C, M, N!** Conceptually, when almost anything moves in the body, it involves muscle tissue. The three types of muscle are as follows:

- Cardiac—found only in the heart, involuntary
  - Lightly striated with one large central nucleus per cell
  - Built-in rhythmicity (will contract on its own without nerve stimulation)
  - Branching pattern
  - Intercalated discs; formed by gap junctions and desmosomes between cells
- Visceral—smooth and involuntary
  - Found in the viscera or guts (e.g., the walls of blood vessels or intestines, etc.)
  - Innervated by the ANS (autonomic nervous system)
- Skeletal—striated and voluntary
  - Must have innervation (nerve supply) to function
  - Allows body movement
  - Connects to bone by tendons, allowing the body to move
  - Innervated by motor somatic neurons

BRIDGE to Chapter 8, Histology if necessary

Muscle Nomenclature and Structure

To understand the structure and function of muscle tissue, we will concentrate on skeletal muscle. Cardiac and visceral muscle are obviously essential for life and do resemble skeletal muscle in many ways, but we'll focus on skeletal muscle to gain conceptual insight while building a foundation of understanding of muscle tissue.

The cell membrane of skeletal muscle cells is called the **sarcolemma**. The cytoplasm of skeletal muscle is referred to as **sarcoplasm,** and the endoplasmic reticulum of skeletal muscle cell is called **sarcoplasmic reticulum**. Notice all three of these specialized names contains the prefix "sarco." Sarco means muscle. So, when you see the prefix "sarco," think muscle.

✔ **CONCEPT!** Sarco as a prefix means muscle.

Skeletal muscle is structurally organized in groupings of three. Figure 11.1 demonstrates that organizational pattern. We'll see the same pattern in nerves!

Protein filaments or myofilaments (*myo* means muscle just like *sarco*) are made up of thick (myosin) and thin (actin) fibers. Multiple myofilaments make up a myofibril. Multiple myofibrils make up a muscle fiber or muscle cell. Muscle cells are surrounded by a delicate connective tissue sheet called the endomysium. Multiple muscle fibers make up a fascicle. Each fascicle is covered by a sheet of dense irregular connective tissue called the perimysium. Multiple fascicles are then wrapped together by a thicker sheet of dense irregular connective tissue called an epimysium. These three connective tissue sheets fuse together to form a tendon. Tendons then cross joints and connect to bone, allowing movement of the body parts. It's easier to grasp this arrangement when looking at Figure 11.1. The following cascade will help as well:

- **Myofilaments**—proteins, myosin and actin
  - **Myofibrils**—bundles of myofilaments

- **Muscle fiber or cell**—bundles of myofibrils, surrounded by the endomysium
  - **Fascicles**—groupings of cells, surrounded by the perimysium
    - **Skeletal muscle**—groupings of fascicles, surrounded by the epimysium
      - Endo + peri + epi = tendon—attaches to bone

BRIDGE to Fig 12.1 to visualize how the nervous system is organized

Motor Units

To function, skeletal muscle must have a nerve supply or nervous innervation.

Figure 12.1 outlines the organization of the nervous system. It shows motor (efferent or outgoing) somatic nerves, providing innervation to voluntary or skeletal muscle. Motor visceral nerves forming the autonomic nervous system (ANS) innervate visceral muscles in the guts located in closed body cavities (review open and closed body cavities). Remember the "R" in REVV CCCAMM!? Repetition is an essential part of learning!

One motor somatic nerve and all the muscle fibers (muscle cells) it innervates is called a "motor unit," as shown in Figure 11.4. The axon of each individual motor somatic neuron splits into numerous branches that innervate individual skeletal muscle fibers. The total number of muscle cells (fibers) each branch innervates determines the size of the motor unit.

This can vary from a small number (less than five) to a large number (several thousand) of muscle fibers (cells). The degree or amount of control in a muscle is determined by the size of these motor units. The muscles that move the eye require a high degree of control and so contain multiple small motor units. The large, powerful muscles of the lower extremity, however, will have many large motor units because they require less control but more force than eye muscles or muscles in the hand..

The area where each branch of the somatic motor axon innervates a muscle cell is called the neuromuscular junction or NMJ. Each NMJ is composed of a synaptic knob at the end of the axon, a synaptic cleft (space), and a motor end plate on the cell membrane of the muscle cell.

Important structures and areas of the NMJ as shown in Figure 11.5:

- Terminal axon of motor somatic neuron; blue
- Motor end plate in sarcolemma with space (light blue) showing the synaptic cleft
- Neurotransmitter (acetylcholine or ACh) produced by protein synthesis of motor neuron

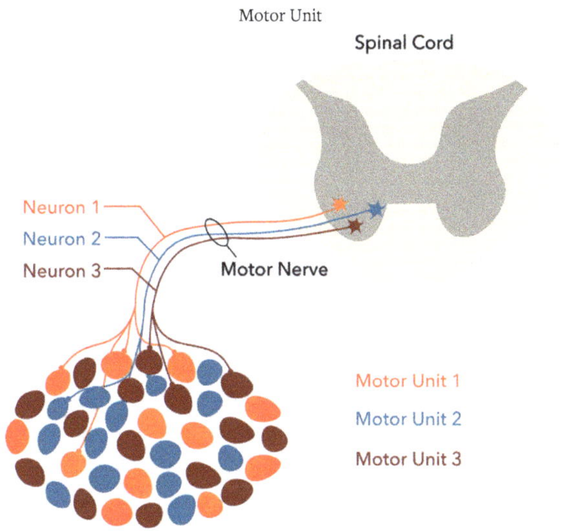

FIG. 11.4 One motor somatic neuron and all the muscle fibers that neuron innervates

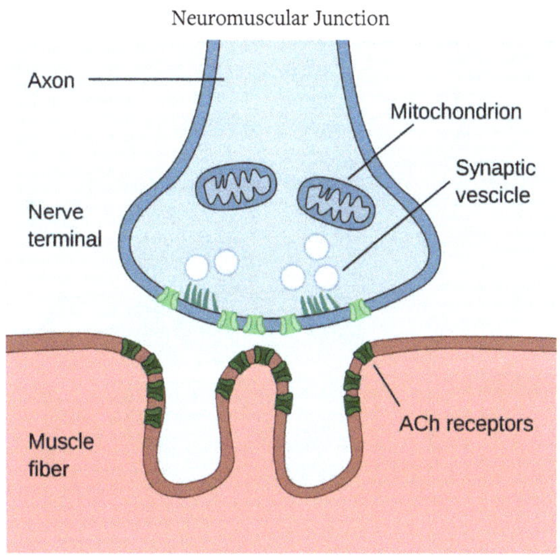

FIG. 11.5 Synaptic site where axon of motor somatic neuron joins the skeletal muscle cell

- NT (neurotransmitters, ACh) packaged in vesicles for release by exocytosis
- Integral proteins (dark green) in sarcolemma (motor end plate) of skeletal muscle
- Integral proteins are ligand or chemically gated protein channels for NT docking
- Mitochondria—muscle cells need tons of energy (ATP)

NOTES

Here is another opportunity to utilize a reference text and/or go online and search for diagrams, discussions, and animations of the NMJ and its function.

Muscle Stimulation and Contraction

Skeletal muscle appears striated because of the arrangement of the actin (thin) and myosin (thick) filaments. This structure is demonstrated in Figure 11.6. The area between the "Z" discs is called a **sarcomere**. It is the contractile unit of skeletal muscle. Introductory A&P students need to know the term "sarcomere" as the contractile unit of skeletal muscle.

An electrical impulse (action potential) travels down the axon of the somatic motor neuron, causing the opening of voltage-gated integral protein calcium channels in the axonal synaptic knob. Calcium ions rush into the synaptic bulb of the motor neuron, causing a series of chemical reactions resulting in the release of a neurotransmitter (NT) from the intracellular vesicles. The neurotransmitter (acetylcholine or ACh in muscle) diffuses across the synaptic space and docks in the ligand-gated integral protein channels. These channels open, allowing sodium ions (Na^{+}) to rush through the sarcolemma and into the sarcoplasm. The rush of sodium down its concentration gradient (channel-medicated facilitated diffusion) causes the sarcolemma to depolarize. This re-establishes the action potential (electrical impulse) in the sarcolemma, leading to muscle contraction.

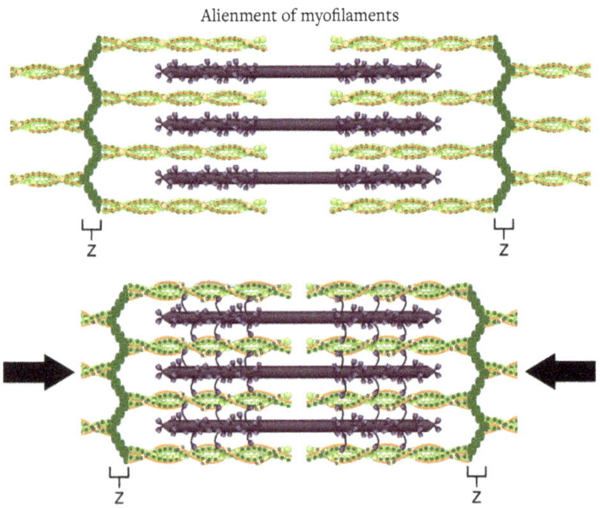

FIG. 11.6 Actin (thin filament) and myosin (thick filament) slide over each other, causing muscle contraction

SUPER simple summary of NMJ function:

- Action Potential (AP) down pre-synaptic somatic motor neuron
- Impulse opens voltage-gated calcium channels

- Ca^{++} ions diffuse into the nerve cell
- Calcium causes chemical reactions releasing ACh through exocytosis
- ACh diffuses across synaptic cleft (down concentration gradient)
- ACh docks with ligand-gated integral proteins in sacrolemma
- Protein gates open, allowing Na^{+} to diffuse into muscle cell
- Post-synaptic neuron depolarizes = AP continues along muscle sarcolemma

Once again there are a cornucopia of animations available online, helping the novice A&P student visualize the structure and function of the NMJ. So, get online and put "Neuromuscular junction" into your browser and have fun gaining additional insight by watching and listening to a variety of animations and video explanations.

✔ **CONCEPT!** Acetylcholine (ACh) is the neurotransmitter found in the NMJ of skeletal muscle.

The electrical impulse (action potential or AP) then travels along the sarcolemma and into microscopic structures in the muscle cell called T-tubules. The action potential stimulates voltage-gated protein calcium channels in the sarcoplasmic reticulum (which stores calcium) to open, allowing for the release of calcium ions into the sarcoplasm. The calcium binds to troponin (a protein that is part of actin), causing the exposure of binding sites on tropomyosin (another protein part of actin). ATP provides energy, causing the head of the thick protein filament myosin to cock. This energized head of myosin will bind to the exposed sites on actin. This forms a cross-bridge, allowing the cocked myosin head to release and pull the actin filament past it. This shortens the length of the sarcomere, resulting in muscle contraction. See Figures 11.7 and 11.8.

Calcium binds to Troponin causing Tropomyosin to rotate exposing binding site for myosin head

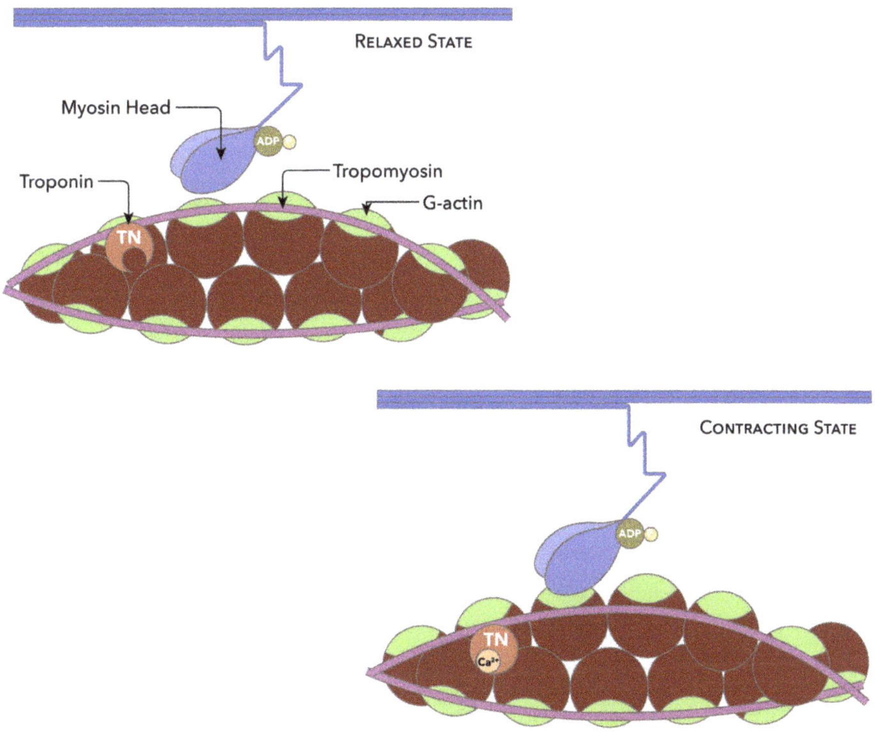

FIG. 11.7 Diagram demonstrating effect of calcium with troponin, causing tropomyosin to expose binding sites for the myosin head

SUPER simple summary of sliding filament theory of muscle contraction:

- AP in sarcolemma travels into T-tubules
- AP opens voltage-gated calcium channels in the SR
- Calcium released from SR into sarcoplasm
- Calcium binds to actin troponin, causing tropomyosin to rotate
- Tropomyosin rotates, exposing binding sites on actin
- ATP provides energy, causing cocking of myosin head
- Myosin head binds to exposed sites on actin, forming cross-bridge
- Myosin head releases causing contraction; power stroke

Here is another opportunity to gain further insight by going online. Just put "sliding filament theory of muscle contraction" in your browser and watch a couple of animations and/or explanations of muscle contraction. This will help you visualize the process and increase your understanding and retention of the concept.

✔ **CONCEPT!** The only active thing a muscle can do is shorten or contract!

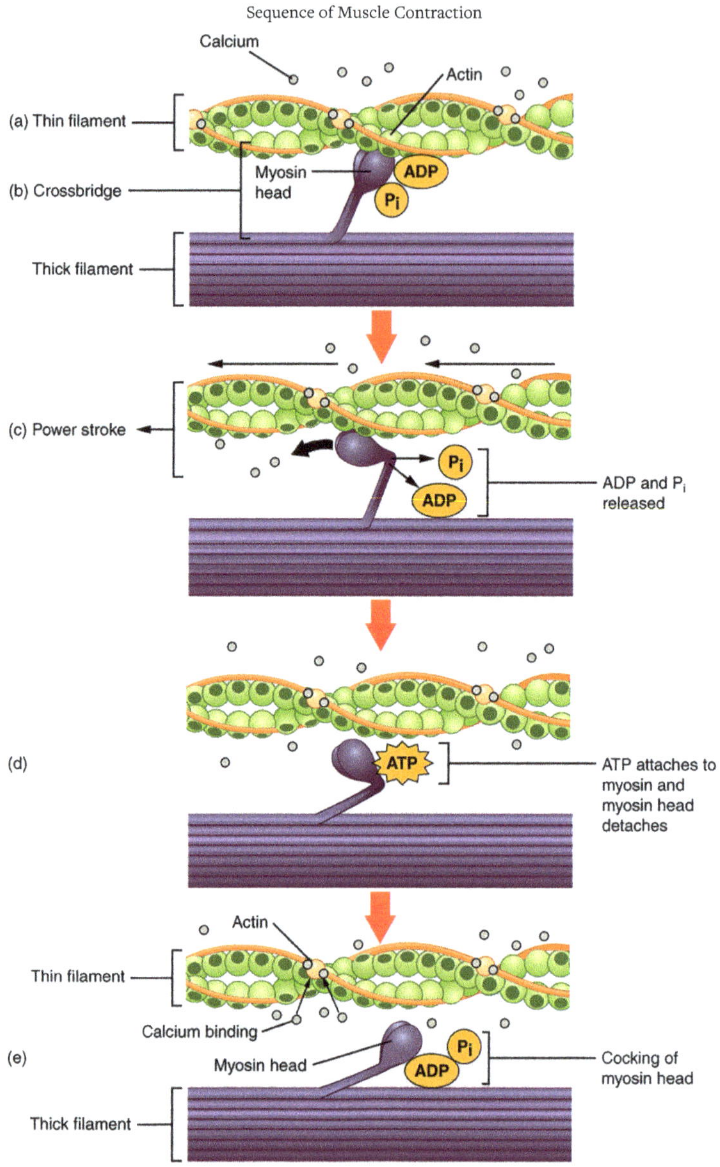

FIG. 11.8 Graphic demonstrating the sequence of events in the sliding filament theory of muscle contraction

Skeletal Muscle Metabolism

Skeletal muscle requires a bunch of energy to function. There are several pathways that provide ATP for muscle contraction. The various means used to meet muscle energy demands are used to classify muscle fibers into three primary types.

- **Type I fibers**—slow oxidative (SO) fibers
  - Contractions are slower and less powerful
  - Contractions occur over longer periods of time
  - ATP primarily supplied by CR (cellular respiration) in muscle mitochondria

- **Type IIa fibers**—fast oxidative (FO) fibers
  - Contractions are fast and powerful
  - ATP primarily supplied by CR

- **Type IIb fibers**—fast glycolytic (FG) fibers
  - Fast glycolytic fibers are called fast anaerobic fibers
  - Contractions are powerful and fast
  - Most common fiber found in skeletal muscles
  - ATP primarily supplied by glycolysis only
  - Contracts for only short periods of time

Types of Muscle Tension

Muscle tension is the force generated when a skeletal muscle is stimulated and contracts. Tension can be described in a variety of ways, depending on the strength and frequency of nervous stimulation.

- **Muscle twitch**—a single brief contraction and relaxation of a muscle
  - Produced by one nerve stimulus
  - Short duration of contraction followed by complete relaxation
- **Multiple motor unit summation**—recruitment of motor units
  - Muscle fibers follow the "all or none" law
    - If nerve stimulus exceeds the threshold—muscle contracts completely
    - If nerve stimulus does not exceed threshold, no contraction
  - Strength of muscle contraction depends on the recruitment of more and more motor units
  - Maximum contraction involves contraction of all motor units in a muscle

- Increasing frequency of stimulation
  - **Treppe**—repeated, more rapid stimulations warm the muscle, which results in a stronger contraction
  - **Wave summation**—incomplete relaxation between stimulation
    - Each contraction summates or builds on top of last contraction
    - Contraction leads to increasing muscle tension
  - **Incomplete tetany**—less time between stimulations
    - Period of relaxation is shorter
    - Contractions fuse together to form a continuous contraction
  - **Tetany**—rapid stimulation
    - Contractions completely fuse
    - Muscle tension is constant; smooth line
  - **Fatigue** or **failure** of muscle—muscle fatigues and then fails

Utilize a traditional reference/resource text and the internet to find helpful graphics or animations of skeletal muscle tension.

Types of Muscle Contraction

Skeletal muscles do not relax, even at complete rest. This resting tension is called "**muscle tone**." The type of movement muscle contraction creates depends on the force generated by the muscle contraction and the amount of resistance to the contraction. There are two types of muscle movement, **isometric** and **isotonic**.

✔ **CONCEPT** Rule of two:

- **Isometric**—no movement during contraction
  - Muscle contracts
  - Muscle tension increases
  - Muscle length stays the same
  - Example—attempting to lift a table that is bolted to the floor
- **Isotonic**—movement during contraction
  - Muscle actively contracts
  - Tension increases causing movement
  - Muscle tone remains the same during movement
  - Isotonic contraction has two subcategories (rule of two)
    - **Concentric**—muscle shortens during contraction
    - **Eccentric**—muscle lengthens during contraction; these are called "negatives" in the gym!

✔ **CONCEPT!** We frequently see one category divided into two subcategories in A&P, just like "isotonic contraction" has two subcategories. One divided into two—rule of two.

NOTES

Cardiac Muscle

Cardiac muscle is only found in the heart. It is striated because it contains sarcomeres, just like skeletal muscle. Cardiac muscle cells branch and are directly connected to each other through gap junctions and desmosomes (review types of cell junctions in Figure 7.4).

BRIDGE to Chapter 7, Fig. 7.4 on page 69 for a graphic of gap junctions and desmosomes

These unique cell junctions are formed the way they are because of what they do. Intercalated discs are unique to cardiac muscle and are caused by the junctions between cardiac muscle cells. The gap junctions allow ions (charge particles such as Na^+ and K^+) to freely move between cardiac cells, allowing the heart to function as a **syncytium** or a unified whole necessary to efficiently pump blood. Desmosomes hold the cardiac cells tightly together, providing a solid unit that is essential since the heart works constantly!

Cardiac muscle is stimulated to contract by a specialized autorhythmic pacemaker, the SA (sinoatrial) node. This feature is responsible for the continual, nonending contractions of the heart necessary for life. (More about the electroconductive system of the heart in chapter 14.)

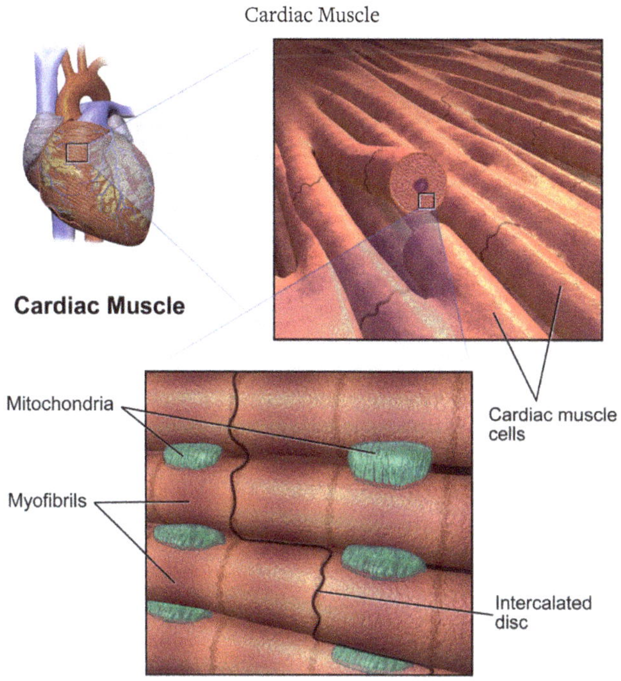

FIG. 11.9 Cardiac muscle, showing mitochondria and intercalated discs

Visceral Muscle

Visceral, smooth, or involuntary muscle is found throughout the body in a variety of structures. Most tubular organs, such as blood vessels or the digestive tract, contain layers of visceral muscle.

Visceral muscle is not arranged in sarcomeres like skeleton and cardiac muscle but does contain actin and myosin filaments. These protein filaments are arranged differently in smooth muscle because of what visceral muscle does.

✔ **CONCEPT!** Things are structured the way they're structured in the human body because of what they do. This is true grossly, at a tissue, or even at a molecular level. Pretty amazing!

Smooth muscle has significant differences compared with skeletal and cardiac muscle in the manner it contracts. However, it still requires the presence of calcium ion (Ca^{++}) ATP, and contracts by the actin and myosin filaments sliding over each other.

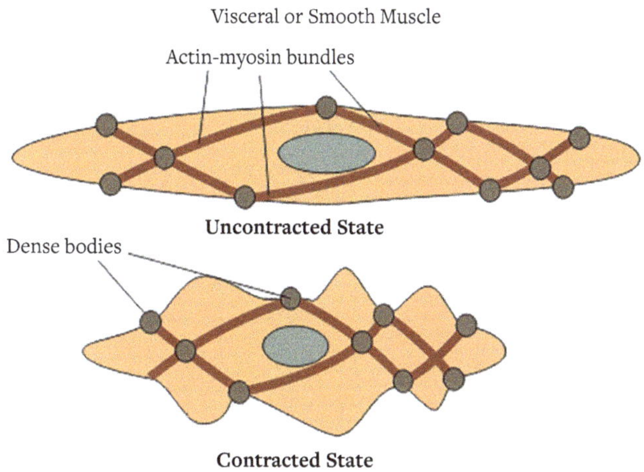

FIG. 11.10 Visceral muscle, demonstrating the unique arrangement of actin–myosin bundles

Review Questions

- What are the three types of muscle in the body?
- What is the only active thing a skeletal muscle (or any muscle cell) can do?
- How does an understanding of the concept of naming muscles help students gain insight into the muscular system?
- What do the terms "synergistic" and "antagonistic" mean?
- What muscle is also called the "tailor's" muscle?
- What muscle is also called the "boxer's" muscle?
- What is the only muscle in the body without any significant boney attachment?
- What division of the nervous system innervates skeletal muscle?
- What is the NMJ, or neuromuscular junction?
- What is the "sliding filament theory" of contraction?
- What are the three connective tissue coverings of skeletal muscle?
- What are myofilaments, myofibrils, and myofibers?
- What is a sarcomere?
- What role does calcium play in muscle contraction?
- What are synergistic muscles?
- What are antagonistic muscles?
- The forearm flexors originate from what boney structure?
- What is the difference between a muscle origin and insertion?

Activities and Coloring (Draw It to Know It)

- Explain the sliding filament theory of muscle contraction to another student.
- Using colored pencils and/or crayons, sketch out (to the best of your ability) a muscle motor unit and label the various structures and areas.
- On a blank sheet of lined paper, write out the seven ways to name skeletal muscles. Then explain each one to someone who doesn't know anatomy.
- List at least one example of a muscle that demonstrates each of the seven naming methods.
- In the blank space provided below draw, color, and label a NMJ. Then explain your drawing to someone else. Remember, vocalizing helps you gain understanding, even if you're talking to your niece or nephew or a cat or dog! They'll love the attention you're giving them!

FIGURE CREDITS

Churn to Learn!

Teach to Learn!

COMMUNICATION AND CONTROL OF THE BODY

CHAPTER TWELVE

The Nervous System

Structure, Function, and Special Senses

TOES (TOPICS OF EMPHASIS)

- Organizational structure of the nervous system
- Basic structure of a neuron
- Classification of neurons—structural and functional
- Neuroglial cells
- Concepts of the CNS, PNS, ANS
- Limbic system
- Cranial nerves
- Spinal nerves
- Categories of sensory receptors
- Skin sensory receptors
- Myelin function
- Action potential—depolarization and repolarization
- Synapse
- Neurotransmitters
- Anatomy of the eye and ear
- Basic concepts of eye function—vision
- Basic concepts of ear function—hearing and balance

Nervous System

The nervous system (NS) is the body's primary communication system. It gathers information from inside and outside the body and transmits that information to the brain and spinal cord (CNS). The CNS (central nervous system) processes that information and responds with outgoing transmissions to effectors like muscles and glands throughout the body.

Our NS does this electrically, very rapidly and with short duration of action. The endocrine system (ES) also communicates inside our body, but it communicates chemically (hormones), relatively (when compared with the NS) slowly and with long duration of action (sometimes decades long). We'll delve into the ES in more detail when we get to chapter 13.

Organization of the NS

Figure 12.1 shows how the human nervous system is structured or organized.

The NS is extremely complex and can be confusing for novice A&P students. *SS A&P* simplifies the NS by eliminating content (there'll be plenty of time for the smaller details later) and focusing on the general concepts of the NS.

✔ **CONCEPT!** The entire NS consists of information going into the CNS (brain and spinal cord); CNS processes that information and sends instructions out of the CNS to effectors (muscles and glands).

Effectors are structures that can respond in some way to stimulation by the NS. These effectors are muscles (cardiac, visceral, and skeletal), and glands. The information coming into the CNS is gathered by a variety of sensory receptors. These receptors are sensitive to temperature, pressure, light touch, position of bones, light, sounds, chemicals (taste and smell), and balance.

✔ **CONCEPT!** In super simple terms, the NS has information going in, it integrates and processes that information, and then it has information (instructions) going out! That's it. That's the whole NS in a simple, conceptual sentence! Understand that basic insight before adding more detail. **Info In—processing—Info out**.

It really is just that simple. That basic concept will form a solid conceptual foundation for understanding the NS. See Figure 12.2.

The structural and functional unit of the NS is the neuron. This is a highly unique cell structured the way it is because of what it does. It's structured to effectively transmit electrical impulses. We'll look at structural and functional differences of neurons later in this chapter, but all neurons have the same basic structure. See Figure 12.3.

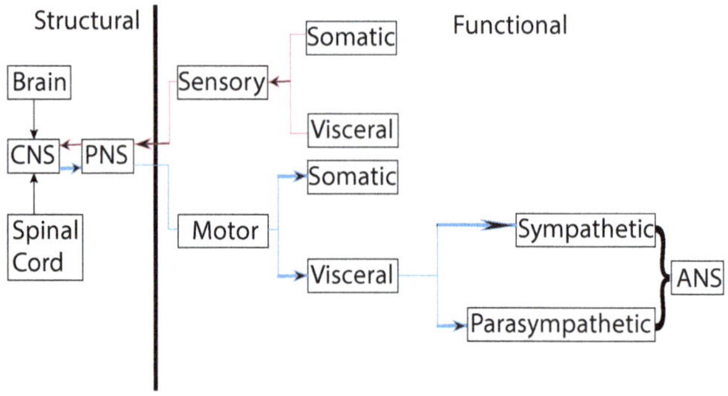

FIG. 12.1 Structural and functional organization of the NS.

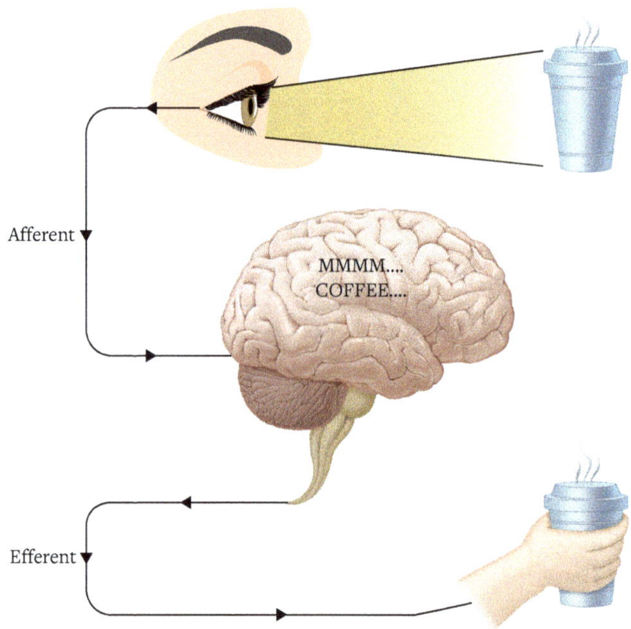

FIG. 12.2 Super simple graphic of the essence of the NS.

Neuron Structure

- **Cell body or soma**—control center of the neuron
- **Dendrites**—relatively short cell processes bring impulses into the cell

  - Nonmyelinated fibers
  - Some neurons have only one dendrite; other neurons have many dendrites
- **Axon**—long process originating from the soma
  - May be myelinated
  - Contacts and stimulates other neurons and effectors such as muscles and glands

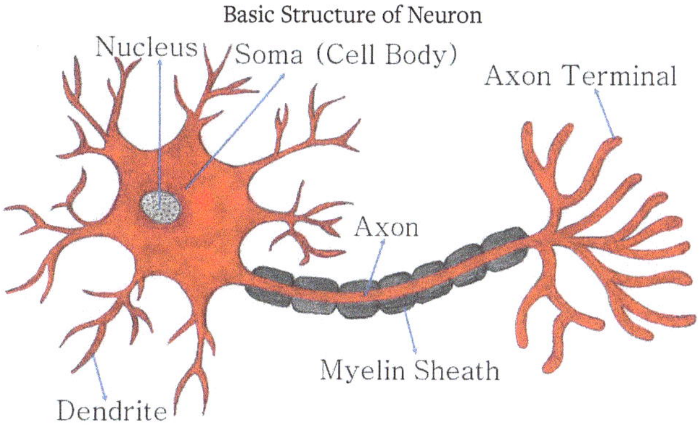

FIG. 12.3 Conceptual structure of a neuron.

Structural Classification of Neurons

- Based on number of processes (dendrites and axons) originating from soma
- Unipolar—only one process; sensory neurons
- Bipolar—two processes; sensory neurons
- Multipolar—three or more processes; motor or association neurons (neurons that connect neurons with other neurons; also called inter-neurons See Figure 12.4.

SUMMARY Chart Comparing STRUCTURAL and FUNCTIONAL Neurons

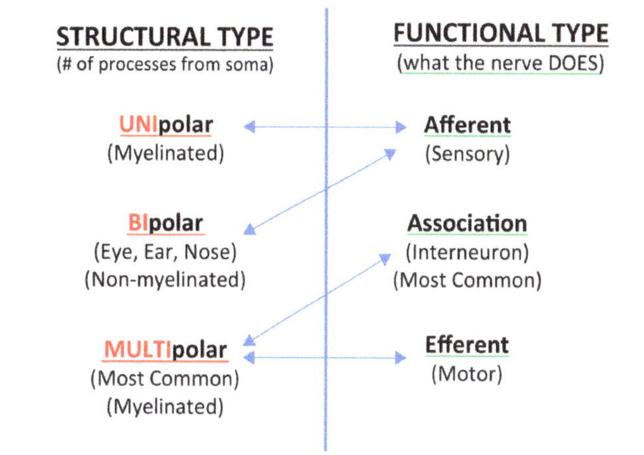

FIG. 12.4 Comparing neuron classifications.

Functional Classification of Neurons

- Afferent or sensory—carry impulses into the CNS (brain and spinal cord)
- Interneurons or association neurons—connect neurons to each other
- Efferent or motor—carry impulses out of the CNS to effectors

Nerves are formed by bundling many neurons together in a grouping of three, just the way skeletal muscles are bundled! It seems the body finds packages of three to be an effective means of wrapping things together.

Notice the connective tissue wrappings surrounding the nerve in Figure 12.5. Just as we saw with the bundling of skeletal muscle, nerves have three connective tissue wrappings.

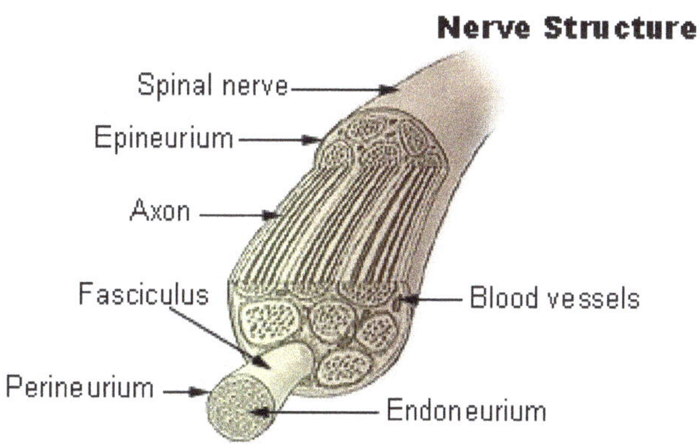

FIG. 12.5 Connective tissue wrapping in a nerve.

- **Endoneurium**—inner wrapping around individual axons of nerve cells
  - Thin, delicate structure
  - Loose areolar connective tissue
- **Perineurium**—secondary layer
  - Dense irregular connective tissue
  - Perineurium forms fascicles—just as in skeletal muscle
  - Nerve is protected and supported by the epineurium

NOTES

Neuroglial Cells—Neuron Support Cells

These cells in the NS do not transmit information but provide vital support for neurons. There are four glial cells found in the CNS and two in the PNS.

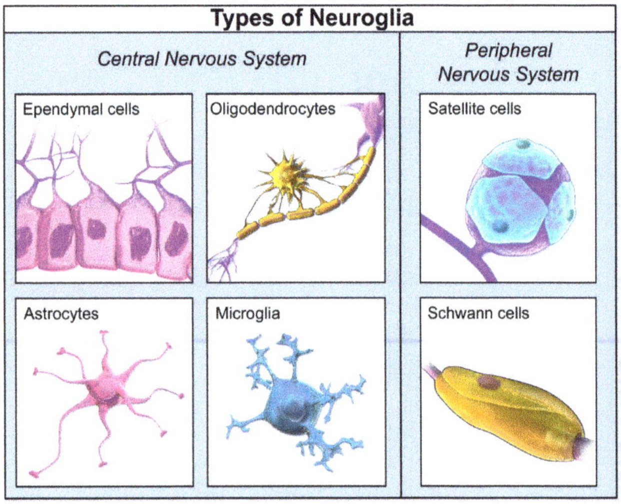

FIG. 12.6 Six neuroglial cells.

- **CNS neuroglial cells**
  - **Astrocytes**—most numerous of the glial cells
    - Helps form the blood-brain barrier (BBB)
    - Star shaped—thus the name "astro" (astro = star, like astronomy!)
  - **Oligodendrocytes**—myelinates axons inside the CNS
    - Myelin insulates axons
      - Increases speed of axon transmission (about fifty times faster)
      - **Nodes of Ranvier**—gaps between myelin sheath segments allow saltatory transmission
      - Produces numerous myelin sheath segments—multiple segments create a myelin sheath
      - Provides myelin sheath segments to multiple axons

- **Ependymal cells**—can act as neuron stem cells; produce new neurons
  - **Choroid plexus**—ependymal cells clumped together in the ventricles of the brain
  - Ventricles of the brain contain ependymal cells, particularly the 3rd ventricle.
  - Central canal of the spinal cord also contain ependymal cells
  - Ependymal cells produce and help to circulate (cilia) cerebral spinal fluid (CSF)
- **Microglial cells**—immune function
  - Cells produce phagocytic action in CNS
  - Infectious agents engulfed by microglial cells

- **PNS neuroglial cells**
  - **Satellite cells**
    - Electrically insulates PNS soma (cell bodies)
    - Regulates nutrient and waste exchange
  - **Neurolemmocytes** (aka Schwann cells)
    - Myelinates and insulates axons in PNS
    - Increases axon transmission rate by as much as fifty times!
  - **Nodes of Ranvier**—gaps between myelin sheath segments allowing **saltatory transmission**

Autonomic Nervous System (ANS)

The ANS or involuntary system involves subconscious control. It can stimulate or inhibit its effectors. For example, the sympathetic division increases heart rate whereas the parasympathetic division decreases heart rate. The autonomic nervous system is automatic!

- sympathetic division creates a "fight or flight" response
- parasympathetic division creates a "rest or digest" response
- sympathetic originates from the thoracolumbar spinal cord
- parasympathetic originates from the craniosacral regions

Central Nervous System (CNS)

The CNS is composed of the brain (averages about three pounds) and spinal cord. These structures, particularly the brain, act as a processing center for the NS. This is where the afferent sensory information is analyzed. Decisions are made, consciously or subconsciously, about appropriate responses to sensory information. The CNS then transmits a response through efferent or motor neurons to effectors like muscles and glands.

We'll examine the gross anatomy of the brain before turning our attention to the spinal cord. The brain is easily the single most complex organ in the body. It is likely to be overwhelming to the novice A&P student, so we will once again dramatically reduce the amount of content and simplify, allowing students to more effectively focus their learning endeavors.

Brain Structure

The brain is composed of four major regions: the cerebrum, diencephalon, brainstem, and cerebellum. The cerebrum is divided into two halves, the left and right cerebral hemispheres. Each hemisphere is subdivided into five functional areas called lobes.

To keep this super simple, students are encouraged to utilize a comprehensive text as a reference or go online and research the following anatomical areas of the brain. These are the areas that are most often emphasized in the typical introductory A&P course.

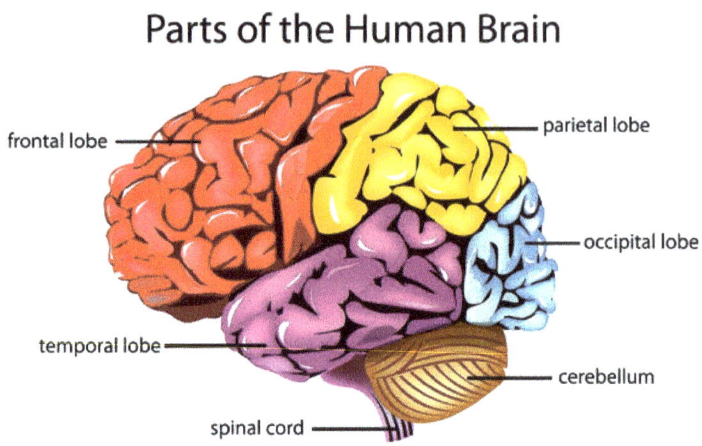

FIG. 12.7 Lateral view of the brain showing major areas.

Anatomical Areas of Emphasis of the Brain

- Protection and support of the brain
  - **Skull**— Boney diploe bone structure
  - **Meninges**—connective tissue layers covering the brain
    - Soft brain tissue separated from the skull by the meninges
    - Pia mater—deepest layer of tissues covering the brain
      - Comprised of loose areolar connective tissue
      - Clings to the brain's surface
    - Arachnoid mater—lies superficial to the pia mater
      - Collagen fibers inside the arachnoid create a delicate web that looks like a spider web, thus the name arachnoid
      - Subarachnoid space located immediately below the arachnoid mater
      - Subarachnoid space contains cerebral spinal fluid (CSF)
    - Dura mater—outermost or most superficial meningeal layer
      - Thick and tough
      - Dense irregular connective tissue

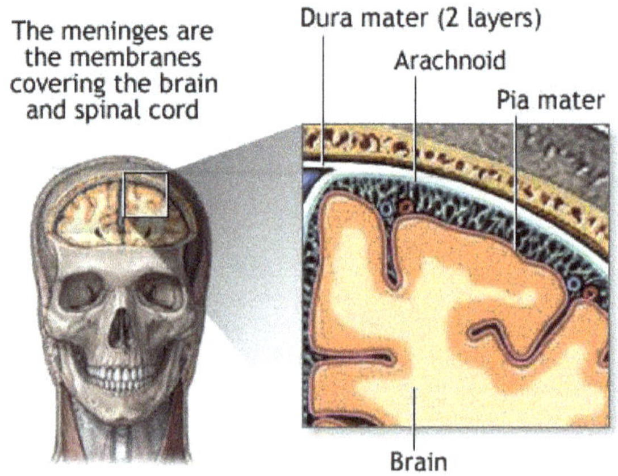

FIG. 12.8 Meninges of the brain.

- Two layers
    - Meningeal—immediately superficial to the arachnoid
    - Periosteal—immediately under the cranium
- **Ventricles**—fluid-filled cavities inside the brain
    - Ependymal cells that produce CSF line the ventricles
    - Cerebral ventricles—left and right
    - Third ventricle
    - Fourth ventricle
    - Cerebral aqueduct connects the third and fourth ventricles

- Lobes (five) of the brain
    - Frontal
    - Parietal
    - Temporal
    - Occipital
    - Insula

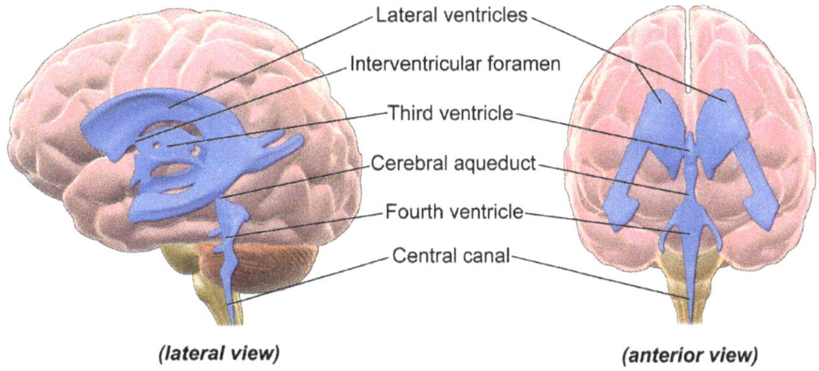

FIG. 12.9 Ventricles of the brain.

Midsagittal Section of the Brain

- Most anatomy labs have brain models showing this plane or perspective. It is a common projection of the inside of the brain. It's also helpful to go online and search for midsagittal images of the human brain.
- Essential structures to know in a midsagittal section of the brain:
    - Lobes—frontal, parietal, and occipital
    - Central sulcus
    - Pre-central gyrus—motor cortex
    - Post-central gyrus—sensory cortex
    - Corpus callosum
    - Thalamus
    - Hypothalamus
    - Cerebrum
    - Midbrain
    - Pons
    - Medulla oblongata
    - Cerebellum

Please label some of the major areas in the midsagittal section of the brain on Figure 12.10.

1. Frontal lobe
2. Thalamus
3. Hypothalamus
4. Diencephalon
5. Pons
6. Medulla oblongata
7. Cerebral aqueduct
8. Cerebellum
9. Spinal cord

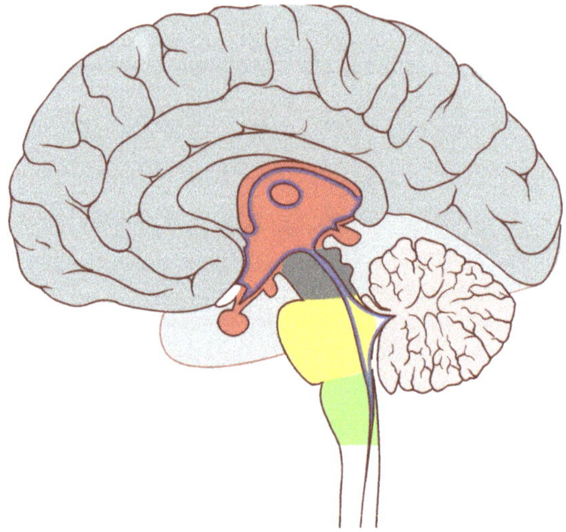

FIG. 12.10 Midsagittal section of the brain.

The brain has gray matter (neuron cell bodies and nonmyelinated fibers) on the outside and white matter (myelinated fibers) mostly on the inside of the brain. The spinal cord has the opposite organization: white matter on the outside and gray matter on the inside.

Two areas of gray matter the novice A&P student needs to learn are the following:

- **Broca's area**—located in the lateral, inferior portion of the left cerebral hemisphere. It is the motor speech area.
- **Wernicke's area**—located in the lateral aspect of the parietal lobe. It is the speech comprehension and understanding area.

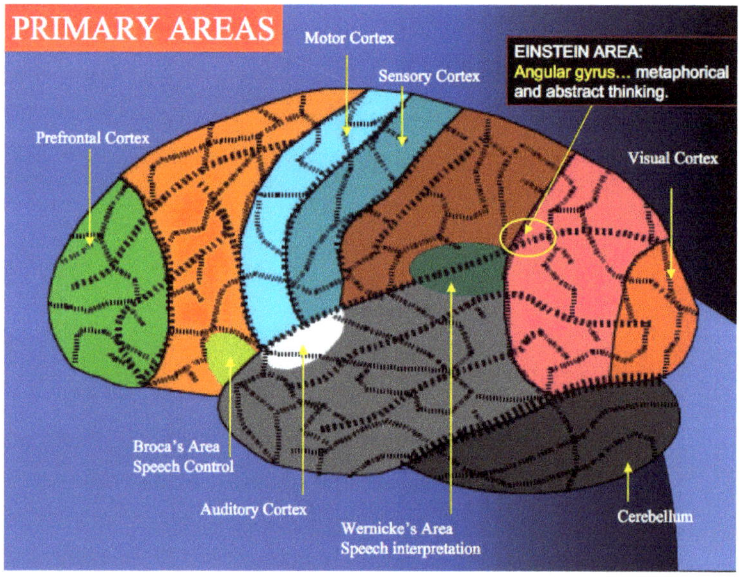

FIG. 12.11 Main functional areas of the brain.

There are clusters of soma (neuron cell bodies) located deep inside the brain. These are called a **nucleus** (singular form) or nuclei (plural form).

Clusters of soma in the PNS, outside of the brain and spinal cord, are called a **ganglion** (singular form) or ganglia (plural form).

The **limbic system** is made up of multiple structures in the cerebrum and diencephalic areas. Collectively they process emotions. The limbic system is sometimes referred to as the emotional brain or the reptilian (primitive) brain. The limbic system structures form a ring around the diencephalon. Although there is some controversy about which structures comprise the limbic system it is commonly thought the following structures are involved:

- Cingulate gyrus
- Parahippocampal gyrus
- Hippocampus
- Amygdaloid body
- Olfactory bulbs, tracts, and cortex
- Fornix

- Several nuclei in the diencephalon
- Mammillary bodies

Notice the limbic system involves the olfactory system or the sense of smell. Smelling is the single strongest sense causing instantaneous emotions or feelings. It is unnecessary for novice students to attempt to memorize the structures of the limbic system. However, it is important to understand the concept of this emotional brain.

Cranial nerves are nerves that arise from the brain or brain stem in the area of the cranium. That is why they're called cranial nerves. Figure 12.12 shows an inferior view of the brain, demonstrating the origin of the cranial nerves.

Figure 12.13 is an Excel chart designed to help you focus on the

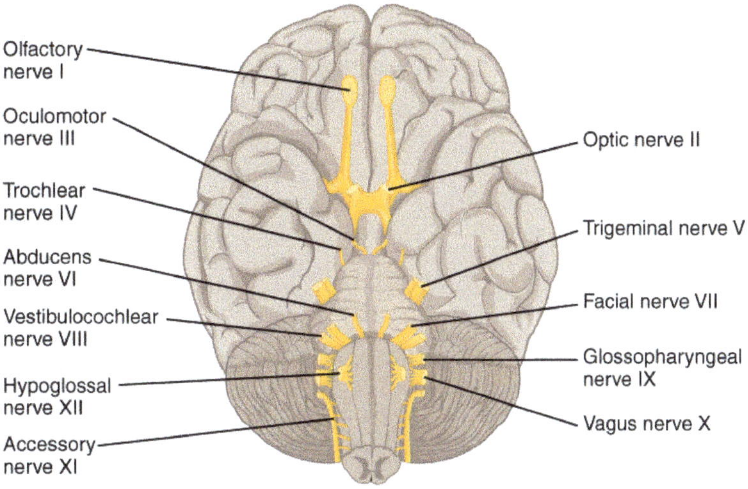

FIG. 12.12 Inferior view of the brain showing origins of cranial nerves.

CRANIAL NERVES Study Guide.....							
#	**Name**	**Type**	**Cranial Passage**	**Target**	**Action/Function**	**Unique ??**	**Clinical Test**
I	Olfactory	S	Cribriform Plate	Olfactory cortex	Smell	bipolar neurons	Smelling
II	Optic	S	Optic Canal	Visual cortex	Vision	bipolar neurons	Snelling chart
III	Oculomotor	M	Superior Orbital Fissure	7/9 eye muscles	Eye movement		Follow the finger
IV	Trochlear	M	Superior Orbital Fissure	Sup. Oblique eye muscle	Eye movement		Follow the finger
V	Trigeminal	B	Opthal-sup. Orbital fissure Max-foramen rotundum Mandib-foramen ovale	Skin of the face and muscles of mastication	Sensation from the skin of the face Muscles of mastication	LARGEST CN	Check sensation of facial skin Chewing
VI	Abducens	M	Superior Orbital Fissure	Lat. Rectus eye muscle	Eye movement		Follow the finger
VII	Facial	B	Stylomastoid foramen	Muscles of facial expression Taste buds	Facial expression Taste		Make faces ! ;o)
VIII	Vestibular-Cochlear	S	Internal Auditory Canal	Auditory cortex Sensory cortex Cerebellum	Hearing and Balance	Contains bipolar neurons	Auditory test Renne and Weber Tests Balance - No eyes
IX	Glossopharyngeal	B	Jugular Foramen	Taste buds Pharyngeal muscles Muscles of mastication	Swallowing Taste Mastication Gag Reflex with Vagus		Drinking Tasting Chewing
X	Vagus	B	Jugular Foramen	Ventral cavity viscera	Visceral sensation Parasympathetic impulses	Longest CN	Uvular deviation Gag reflex
XI	Spinal Accessory	M	Jugular Foramen	Muscles of head and neck	Head/neck movement	Only CN with origin in the cervical SC	Shoulder Shrug Head turn against hand resistance
XII	Hypoglossal	M	Hypoglossal canal	Muscles of the tongue	Tongue movement		Stick out tongue Push tongue against cheek
						NHKraus2010	

FIG. 12.13 Need-to-know information about cranial nerves.

essential things to know about the 12 cranial nerves (CNs). This will be one place where students will have to do some good, old-fashioned memorization. Even a "SUPER simple" approach can't eliminate all memorization, just reduce it a bunch. After all, memorizing is an essential part of learning anything! REVV CCCAMM!

Students need to know the number (Roman numeral), name, cranial passage, target, function, and uniqueness of the cranial nerves. Again, this is one of those places in A&P where the student will need to do the repetition necessary to memorize the CNs.

However, there are memory tricks you can apply to make it an easier task. Some traditional textbooks will provide mnemonics, which can be helpful. Or you can be creative and make up your own mnemonic to help you retain the important information regarding CNs. There are also a variety of mnemonics available online. So, have some fun while you're learning a ton about the CNs.

NOTES

Upper and Lower Motor Neurons

Motor neurons located in the pre-central gyrus of the cerebral cortex or in the brain stem are called upper motor neurons, whereas motor neurons located in the ventral horn of the spinal cord are called lower motor neurons. Outgoing efferent impulses originate in the upper motor neurons. They travel to a lower motor neuron, which then stimulates an action in an effector, a muscle or gland.

Spinal Cord and Spinal Nerves

The spinal cord (SC) is a vital link between the brain and the rest of the body. But the spinal cord does have some functional independence from the brain, like reflex arcs.

- Structure—the SC is like the vertebral column in its anatomical organization
  - Four different sections
    - Cervical
    - Thoracic
    - Lumbar
    - Sacral
  - Because students are already familiar with these names, it's easy to deduce where each section of the SC is located.
  - On cross section the SC has a classic butterfly-type of pattern. Figure 12.14 shows this.

- Notice that the spinal cord contains white matter (myelinated fibers) on the outside and gray matter (soma and nonmyelinated fibers) on the inside of the cord, whereas, the brain is just the opposite arrangement: white matter mostly on the inside and gray matter on the outside.
- Dorsal horns of the spinal cord mostly involve sensory (incoming) function.
- Ventral horns mostly involve motor (outgoing) function.
- Spinal cord nerve roots will join to form spinal nerves.
- Dorsal roots are sensory (incoming info).
- Ventral roots are motor (outgoing info).
- Spinal nerves are mixed nerves, combining sensory and motor functions.

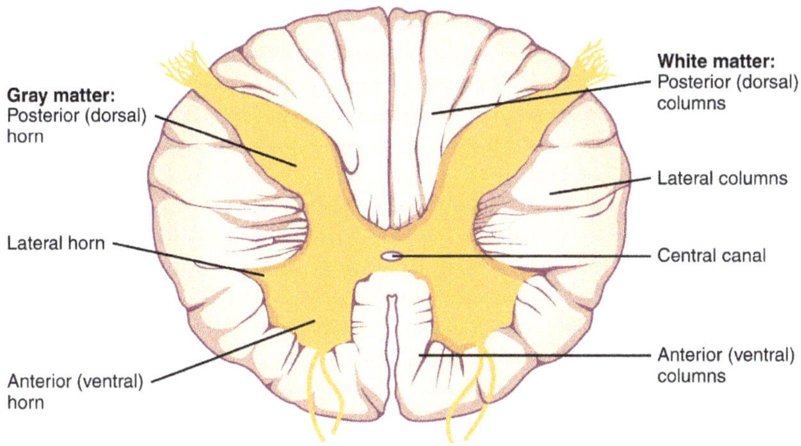

FIG. 12.14 Cross section of the spinal cord.

NOTES

- Spinal nerves and segments
  - Cervical portion of SC has eight segments and spinal nerves.
  - Thoracic segment has twelve segments and spinal nerves.
  - Lumbar segment has five segments and spinal nerves.
  - Sacral segment has five segments and spinal nerves.
  - Coccygeal segment has one segment and one spinal nerve, although some consider this coccygeal segment to be the most inferior part of the sacral portion.

- Cervical enlargement is an enlarged region of the inferior portion of the cervical cord, caused by an increase in neurons providing innervation to the upper extremity.
- Lumbosacral enlargement is an enlarged region of the mid-lumbar portion of the SC, caused by an increase in the soma providing innervation to the lower extremity
- Be aware: The different segments or portions of the SC do not match up exactly with the vertebrae of the same name.
- SC proper ends at L1 vertebrae; the ending is called the **conus medullaris.**
- Spinal nerves continue to travel inferiorly from the ending of the SC at L1.
  - Collectively these spinal nerves are called the **cauda equina.**
  - Cauda equina means "horse's tail" (grossly it looks like a horse's tail!).
  - **Filum terminale** is a thin film of the pia mater of the SC, which continues inferiorly helping to anchor the SC to the coccyx.

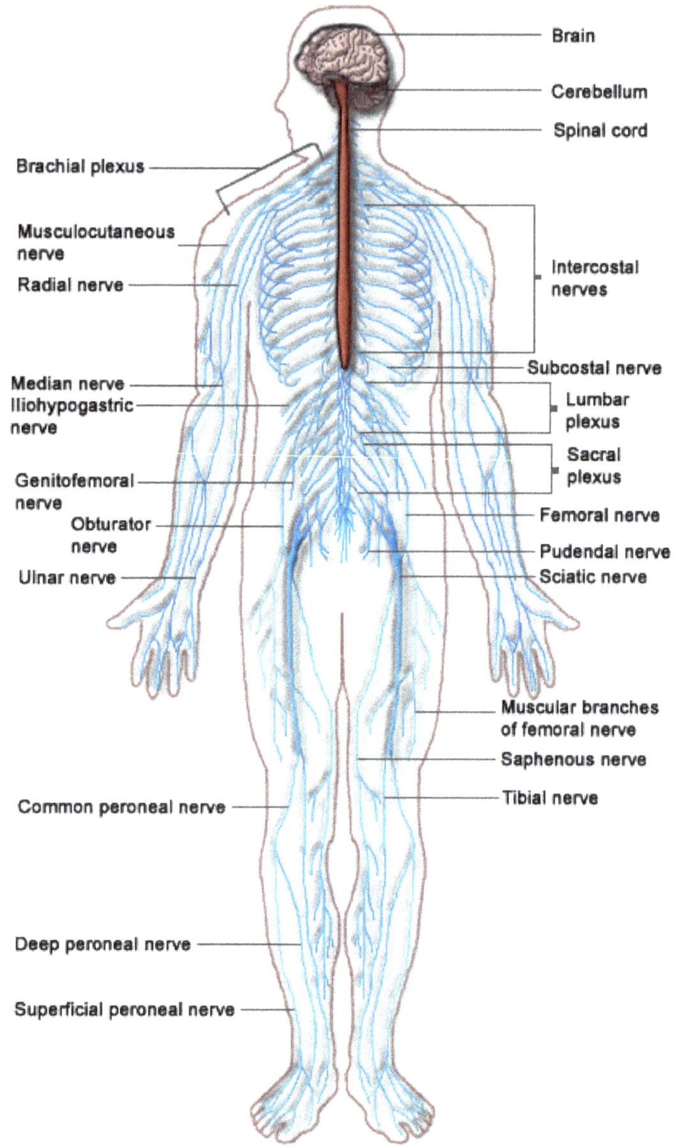

FIG. 12.15 Major spinal nerves.

- Spinal nerves join and mix in areas called plexi and then form a multitude of peripheral nerves, innervating distal areas of the body. Spinal nerves are mixed nerves, containing both sensory and motor fibers.
  - Cervical plexus—involves spinal nerve roots C1 through C4, which serve muscles and areas of the skin in the head and neck
  - Brachial plexus—involves spinal nerve roots C5 through T1, which serve muscles and cutaneous areas in the upper extremity
  - Lumbar plexus—involves spinal nerve roots L1 through L3, which serve muscles and cutaneous areas in the abdomen, external genitalia, and the buttocks
  - Sacral plexus—involves spinal nerve roots L4 through S4, which serve muscles and cutaneous areas in the lower extremity

✔ **CONCEPT!** Understand the idea of a plexus rather than struggling to memorize all the specific details of a nerve plexus. Each plexus is a network of interweaving spinal nerves, which then split into the multiple peripheral nerves.

✔ **CONCEPT!** Some of the peripheral nerves are named for the body region they occupy. So, if a student knows the basic regions of the body (Chapter 4) it's easy to understand where you'll find the femoral nerve, the brachial nerve, the radial, or the ulnar nerve. Once you grasp the concept it eliminates the need for so much rote memorization.

BRIDGE to Chapter 4 to review the body regions in Figure 4.6 on page 36

Reflex Arc

Rapid, involuntary responses to stimuli are called reflexes. Quickly pulling your hand away from a hot stove burner is an example of a reflex. The knee-jerk reflex frequently tested by a doctor or nurse practitioner with a reflex hammer is another example. Figure 12.16 demonstrates how that reflex works.

The components of a reflex arc are as follows:

- Stimulus—initiates the reflex
- Response—rapid
- Preprogrammed response—reflex is the same each time
- Involuntary response—no conscious awareness required

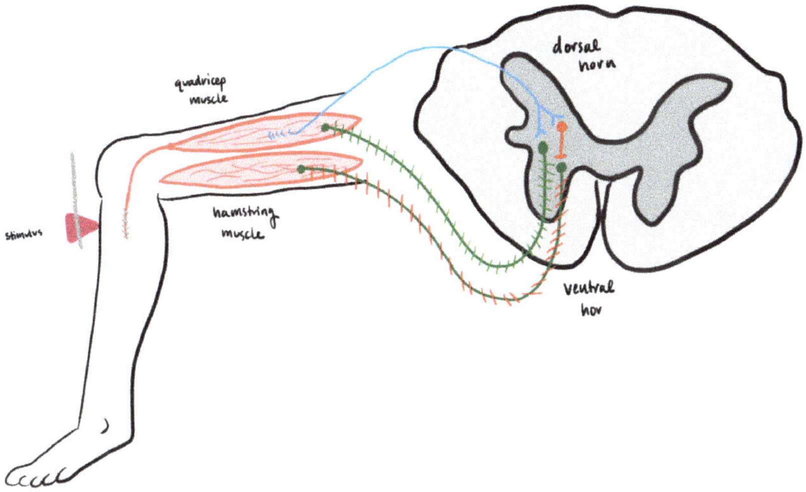

FIG. 12.16 Knee-jerk reflex arc.

NOTES

Special Senses—Eye and Ear

The special senses are vision, hearing, smell, taste, and equilibrium. They are a vital aspect of A&P. However, at this basic, introductory level, we'll only focus briefly on the eye and the ear.

Eye

The eye is a complex nerve receptor sensitive to light. It allows the body to determine light and dark. For foundational purposes, Figure 12.17 is a good image of the basic structure of the human eye.

In addition, the introductory student needs to learn about the lacrimal gland and lacrimal duct. The gland (located laterally) produces tears flowing from the lateral aspect of the eye to the medial aspect, next to the nose. Tears then drain down the nasolacrimal duct into the nasal cavity. That's why our nose runs when we cry. Tears overflow onto the cheeks and drain down into the nose, so we need to grab a tissue.

Jump back to Figure 11.3 on page 138 showing extraocular eye muscles. Remember these are striated muscles under voluntary control, which move the eyeball.

The eye is a spherical structure occupying the orbit of the skull. It is composed of three general layers:

- Fibrous tunic—sclera and cornea
- Vascular tunic—iris, ciliary body, and choroid
- Retina—pigmented layer and the neural layer containing rods and cones

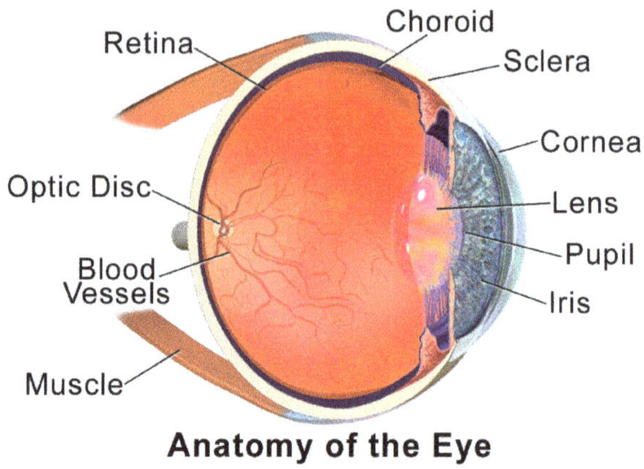

FIG. 12.17 Lateral or sagittal view of the eye with major structures.

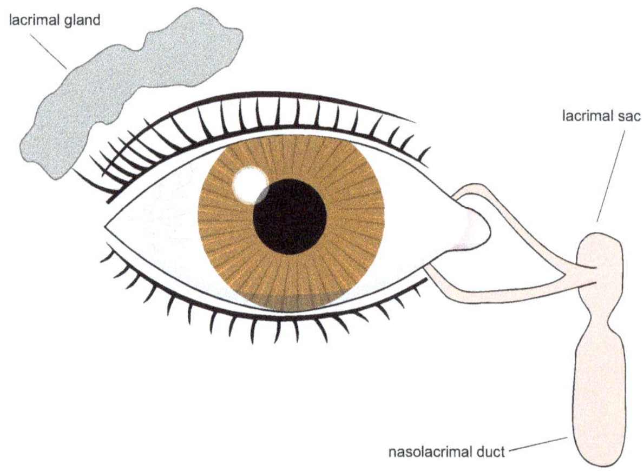

FIG. 12.18 Lacrimal gland, sac, and duct.

BRIDGE to Chapter 11, Figure 11.3 on page 138 to review the extraocular eye muscles

The introductory A&P student needs to know the location and function of the following structures or areas of the eye. These are listed from the anterior to posterior aspect of the eye.

- Cornea
- Anterior cavity—contains aqueous humor
  - Anterior chamber
  - Posterior chamber
- Iris
- Pupil—not a structure but an opening in the iris
- Lens

- Ciliary body—contains the ciliary muscle (What type of muscle is this?)
- Lens
- Suspensory ligaments—connect the lens to the ciliary body
- Posterior cavity—contains vitreous humor
- Retina
- Macula lutea containing the fovea centralis
- Optic disc—cranial nerve II

Ear

The ear is the organ that detects not only sound but movement of the head. Stimuli are transduced (converted) into nerve signals and carried in CN VIII (vestibulocochlear nerve) to the brain, producing the sensations of hearing and equilibrium. Figure 12.19 demonstrates the major parts of the ear apparatus. Notice there are three distinct anatomic regions of the ear: external, middle, and inner. The external ear is located outside the skull. That's why it's called external!

✔ **CONCEPT!** Remember, most names in A&P are very logical. The names tell you something about the structure. Use this concept to your advantage to minimize memorization. The external ear is called external because it's located outside the skull. The inner ear in located inside the skull, and the middle ear is located in between the external and inner ear. It's all very logical and makes sense.

The novice A&P student needs to know the following structures/areas of the ear and their function. These are listed from the external to the internal aspects of the ear:

- Auricle or pinna
- External auditory canal
- Tympanic membrane
- Tympanic cavity
- Malleus
- Incus
- Stapes
- Oval window
- Cochlea
- Vestibule—saccule and utricle
- Semicircular canals
- Round window
- Eustachian tube—pharyngotympanic tube
- CN VIII

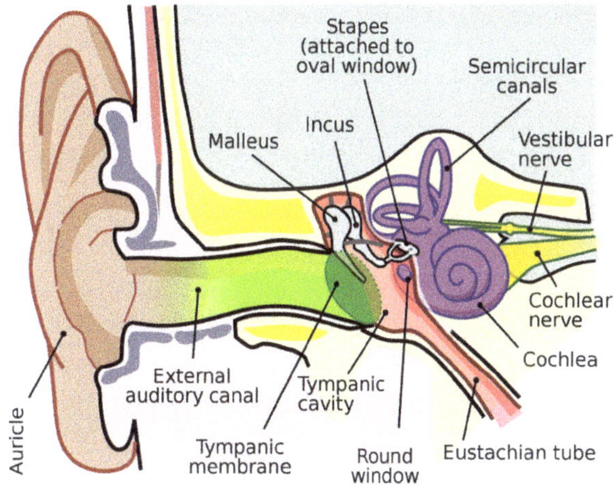

FIG. 12.19 Major structures of the ear.

NOTES

Nervous System Physiology—Electrical Communication

H2O

The nervous system (NS) is the body's primary communication system. The H_2O sheet shows five Cs as the function for the NS. These stand for communication, control, command, coordinate, and consciousness!

✔ **CONCEPT!** The NS provides CCCCC!

- Communication
- Control
- Command
- Coordination
- Consciousness

The NS is organized structurally and functionally, as shown in Figure 12.1.

It is essential for the novice student to understand this organizational diagram. The NS involves two (rule of two) categories, structural and functional. The structural category has two major divisions: the CNS (central nervous system) and the PNS (peripheral nervous system) (rule of two again). The CNS has two parts, the brain and the spinal cord. If a NS structure is not brain or spinal cord, then it must be a part of the PNS!

✔ **CONCEPT** Rule of two: One area or division divided into two subcategories

The PNS also has two divisions, sensory (afferent) and motor (efferent). Sensory brings impulses into the CNS from the organs of the body or sensory receptors (eyes, ears, skin, etc.), which sense environmental changes. The sensory and motor divisions each have two (surprise!) divisions, somatic and visceral. The motor visceral division has, as you would guess, two subdivisions, the sympathetic and parasympathetic, making up the ANS, or autonomic nervous system.

Make sure you understand this organization. Practice writing the organizational flow chart, Figure 12.1, on a blank sheet of paper from memory.

The structural and functional unit of the NS is the **neuron**. The neuron is a unique cell that transmits information via electrical impulses (action potentials).

The dendrites of the neuron receive stimulation from receptor cells or other neurons and transmit impulses in only one direction, dendrites, soma (cell body), axon hillock, and axon. **The neuron is a one-way highway**! Neurons function in one of three manners. They are either sensory (afferent or incoming), motor (efferent or outgoing), or interneurons (association neurons that are neurons hooking neurons together). The chart in Figure 12.4 shows the relationship between structural and functional neurons.

The number of processes originating from the cell body or soma determines the structural type of neuron. Notice bipolar neurons are never myelinated, whereas unipolar and multipolar are myelinated. Functionally, neurons are either sensory, motor, or interneurons (association neurons). The association neurons connect neurons, usually a sensory neuron to a motor neuron.

NOTES

One of the most important concepts to grasp regarding the NS is the **AP**, or **action potential**. This is the process that allows nerves to transmit an electrical impulse into and out of the CNS.

There is an electrical charge difference (electrical gradient) across the resting neuron cell membrane. The inside of the axon is more negatively charged than the outside of the membrane. This is called the **RMP**, or resting membrane potential (See Figure 12.20). Typically, the RMP is a negative 70 millivolts (–70 mV). The RMP is created by an imbalance in the concentration of sodium ions outside the membrane and potassium ions inside the cell. In addition, the presence of large protein anions trapped inside the cell, because of their large size, helps create and maintain the RMP. The sodium/potassium pump is a factor maintaining the RMP by pumping more sodium out of the cell than potassium into the cell.

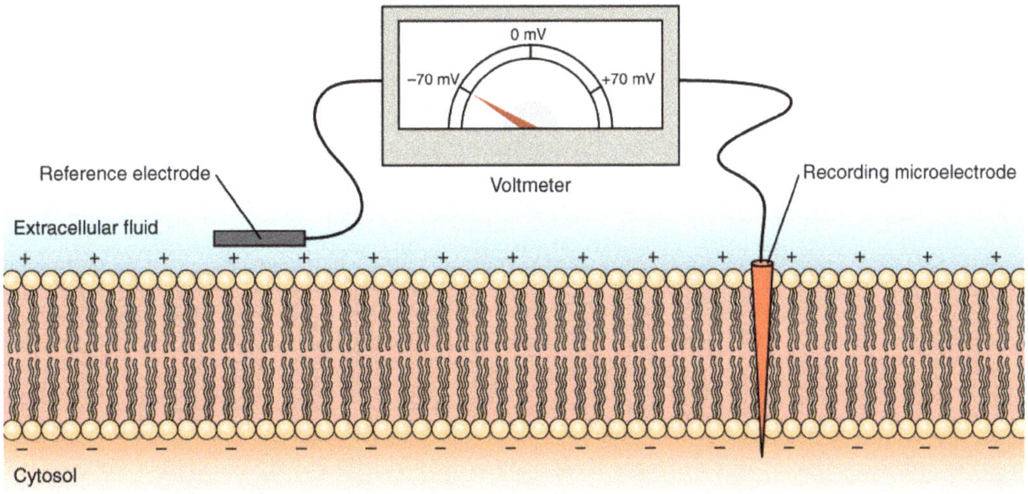

FIG. 12.20 Demonstration of RMP in an axon.

Stimulation of a neuron begins with the establishment of **graded potentials** (GP) in the receptive area of the neuron (dendrites and soma). GPs are relatively (compared with APs) small, short-lived, local fluctuations of the RMP caused by the movement of small amounts of ions across the cell membrane when receptors are stimulated. Unless numerous GPs summate and reach the threshold level (–55 mV), no AP is initiated. GPs that occur in postsynaptic neurons as the result of neurotransmitter stimulation can create either **EPSP** (excitatory postsynaptic potential) or **IPSP** (inhibitory postsynaptic potential). EPSP means the RMP is reduced in the postsynaptic neuron, making it

BRIDG
to Figure 12.3 on page 155 to review neuron structure

more likely to reach threshold and fire off an AP. IPSP means the RMP is increased in the postsynaptic neuron, making it less likely to reach threshold and fire off an AP.

APs fire off when the GPs reduce the RMP to the **threshold level** of −55 mV. When the threshold is reached, voltage-gated protein channels open, allowing sodium ions to rush down their concentration gradient from extracellular space into the cell. This depolarizes the axon membrane voltage potential, flipping it from −70 mV to +30 mV. The AP flows to adjacent areas down the axon, causing additional protein channels to open. Remember, this only occurs in one direction, away from the soma or neuron cell body. The AP impulse flows down the axon to the axon terminal synapse, where it opens the integral calcium protein channels, causing the release of a neurotransmitter (like ACh in the NMJ) resulting in a reaction in an effector or another neuron.

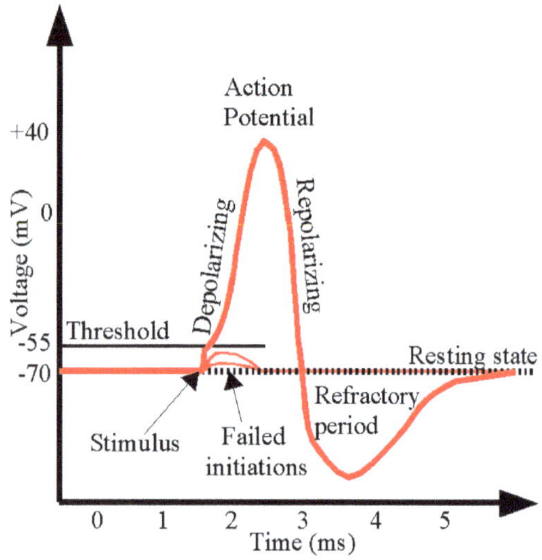

FIG. 12.21 Graph of typical AP.

Before the axon can fire off another AP, the membrane must return to its RMP. This occurs when the sodium channels close and potassium channels open, allowing potassium to rush from inside the cell to outside the cell. This returns the RMP to −70 mV and the axon is ready to transmit another AP.

Because the potassium channels close slowly there is an overshoot of the flow of potassium ions, resulting in a period of **hyperpolarization**.

During the AP there are two refractory periods. The **absolute refractory period** occurs during the AP, both depolarization and repolarization. During this time, the nerve will not fire off another AP no matter how strong the stimulus. The area of hyperpolarization creates a **relative refractory period** where another AP may fire off but only with a strong stimulus.

Some axons are nonmyelinated and transmit APs at a slower rate than those axons covered with myelin. Remember that **oligodendrocytes** produce myelin sheath segments in the CNS, whereas **Schwann cells** (**neurolemmocytes**) produce myelin sheaths in the PNS. Myelin sheaths act much like insulation on an electric cord. The myelin sheath increases the speed of APs transmission approximately fifty times. Conduction in a nonmyelinated axon is called continuous, whereas conduction in a myelinated axon is called saltatory. In **saltatory conduction**, the impulse jumps or "leap frogs" from one node of Ranvier to the next node (see Figure 12.3).

NOTES

Although the human has billions of neurons, what provides the marvelous capabilities of the NS are **synapses**, neurons connecting (hooking up) to other neurons to form neuronets. It's the process of building and restructuriing synapses that allow humans to walk, talk, and learn. A neuronal synapse is just like the neuromuscular junction (NMJ), discussed in the previous chapter (chapter 11), except it's a junction between two neurons rather than a neuron and a muscle cell.

BRIDGE to Figure 11.5 on page 140 to refresh your understanding of the NMJ

A synapse is made up of a presynaptic neuron, a synaptic cleft (space), and a postsynaptic neuron. The membranes of the neurons contain integral proteins acting as voltage-gated or ligand-gated channels, facilitating the movement of ions into and out of the cells.

As in the NMJ, an electrical impulse (AP) travels down an axon, arriving at the axonal knob or bulb. The AP stimulates voltage-gated channels to open allowing calcium ions to rush down their concentration gradient into the cell. Inside the neuron, these calcium ions create a chemical cascade, resulting in the release of a neurotransmitter (like ACh) produced by the neuron through protein synthesis and packaged in vesicles. The neurotransmitter is released through exocytosis into the synaptic cleft. It diffuses across the synaptic cleft and docks with ligand-gated protein channels in the membrane of the postsynaptic neuron. These channels open, allowing sodium ions to rush down their concentration gradient, causing depolarization of the postsynaptic neuron.

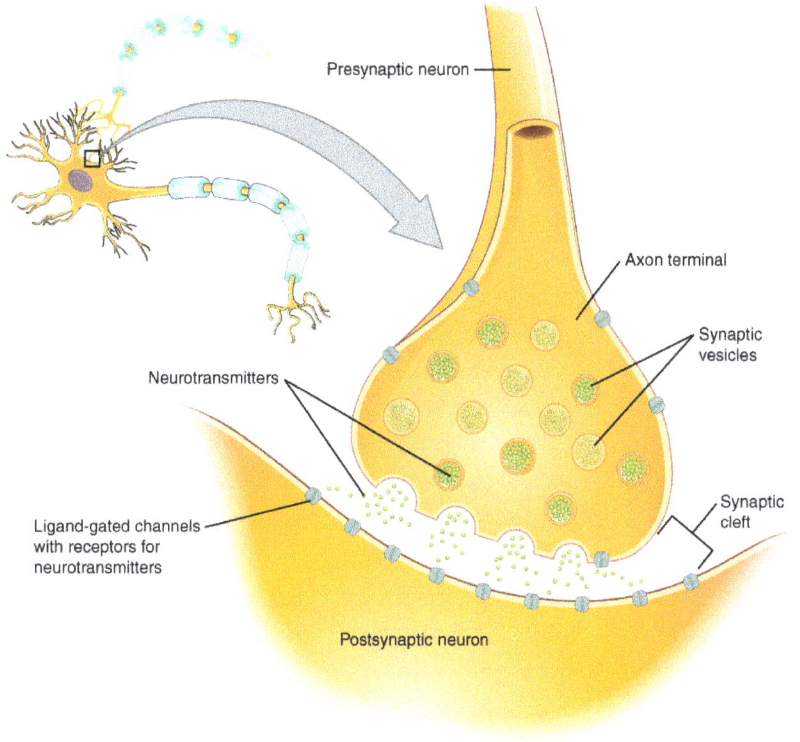

FIG. 12.22 Neuronal chemical synapse.

SUPER simple summary of synaptic function:

- AP travels down the presynaptic neuron
- Impulse opens voltage-gated membrane proteins
- Ca^{++} ions diffuse into cell
- Calcium causes chemical reactions releasing NT (ACh or other NT)
- NT diffuses across synaptic cleft
- NT docks with ligand-gated membrane proteins in postsynaptic neuron
- Proteins open, allowing Na+ to diffuse into cell
- Post-synaptic neuron depolarizes = AP continues

Check out a reference text or online resources (e.g. "Crash Course") for images, discussions, and animations of synaptic function.

Novice A&P students need to be aware there are different types of neurotransmitters that function at synapses. The neurotransmitter at the NMJ is ACh. It has a unique chemical structure, so it is placed in its own category, but to simplify we'll consider

ACh to be a protein-like molecule. ACh performs a variety of functions in numerous synapses in skeletal, cardiac, and visceral muscle, as well as other organs and glands. Most neurotransmitters are protein molecules structured to interact with specific integral protein receptors in cell membranes.

Classification of neurotransmitters:

- Acetylcholine (ACh)—unique chemical structure; contains nitrogen so "protein like"
- Biogenic amines—derived from amino acids
- Amino acids (e.g., glutamate, aspartate, GABA)
- Neuropeptides—chains of amino acids

NOTES

Sensory Nervous System and Special Senses—Eye and Ear

The body has five special senses, but at a novice level we're going to focus on the physiology of three of those senses: vision, hearing, and balance, the eye and the ear. Sure, smell, taste, and touch are essential. However, it can be argued the eye and ear have more direct clinical significance. And at an introductory level of A&P it makes sense to limit our content and spend more time concentrating on less material. This will foster insight and retention.

Categories of Sensory Receptors

Before we discuss the eye and the ear, it's important to look at the types of sensory receptors that initiate the input of sensory information into the CNS. These are classified based on what type of energy stimulates the specific receptor.

- **Chemoreceptors**—respond to chemicals in the environment
  - Smell
  - Taste
- **Thermoreceptors**—respond to either heat or cold (separate receptors for each)
- **Mechanoreceptors**—respond to mechanical stimuli
  - Touch and pressure in skin
  - Hearing and equilibrium in inner ear

- **Photoreceptors**—retina of eye; sensitive to light
- **Nociceptors**—pain
- **Proprioceptors**—stretch receptors in muscle, joints, tendons, and ligaments
  - Muscle spindles
  - Golgi tendon organs
- **Cutaneous receptors**—receptors in skin for touch, pain, temperature, and so on
  - Pain and temperature—naked dendrites
  - Tactile (Meissner's) corpuscles
  - Lamellar (Pacinian) corpuscles

As always in A&P, an understanding of the anatomy of the eye helps when learning about the physiology or function of the eye. Figure 12.23 is similar to Figure 12.16 but with more detail.

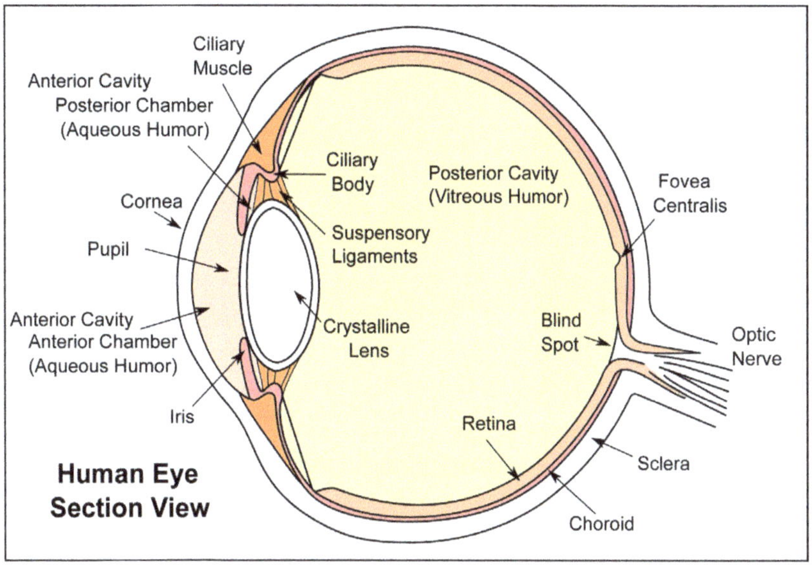

FIG. 12.23 Basic structure of the eye. Similar to Figure 12.16 but with more detail.

Eye Physiology

Here are the foundational physiological concepts and material the novice A&P student needs to know about the eye:

✔ **CONCEPT!** The eye turns light into pictures in our head!

- **Focus light on the retina**
  - Refraction is the bending of light passing from one medium to another
  - Cornea does not change shape and so provides a constant amount of refraction
  - Cornea and lens are transparent because
    - Avascular (no blood vessels)
    - Cells do not contain any intracellular organelles
  - Cells contain protein—crystallin
  - Cornea is major light refractor—not the lens which accommodates
- **Lens is adjustable—accommodation**
  - Process keeps light focused on retina when changing from close to distant vision (like reading and then looking out the window)
  - Contraction and relaxation of ciliary muscle changes lens shape

- **Refractive problems create visual acuity**
  - Myopia = nearsightedness
  - Hyperopia = farsightedness
  - Astigmatism = cornea and/or lens not curved symmetrically, causing light to scatter, creating blurry vision

✔ **CONCEPT!** The **cornea** causes refraction of light. It is most responsible for focusing light on the retina. That's why contact lenses correct vision. Think about it! Where does a person place their contact lens? On the cornea!

Changing Light into Action Potentials

- Retina—structure that changes light to APs
- Forward extension of the brain

Neural layers face outward into the vitreous humor toward the light coming in through the cornea, pupil, and lens.

There are two types of highly specialized photoreceptor cells located deep to other neural layers. Light must pass through those other neural layers to reach the photoreceptor cells:

- Rods—provide black and white vision at low light intensity
- Cones—provide sharper images and color images at higher light intensity; three types:
  - Blue
  - Green
  - Red

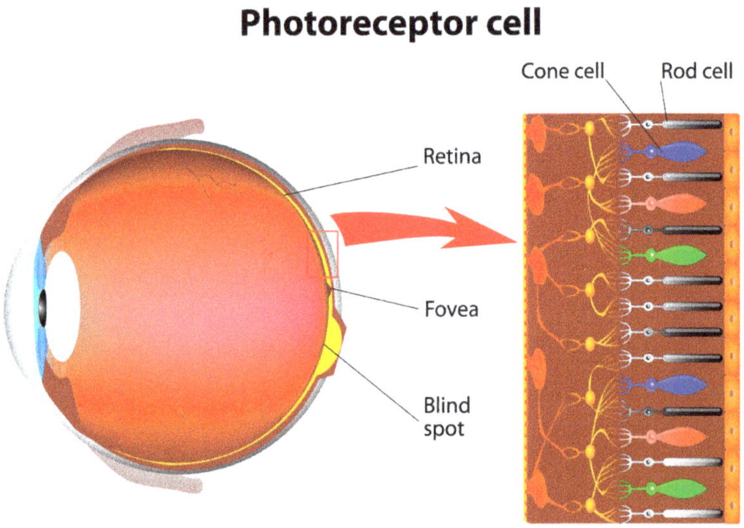

FIG. 12.24 Details of photoreceptor cells in the retina.

Effect of light hitting the retina:

- Light hitting retina causes the stimulation of the rods and cones
- Different types of cones are stimulated by different wavelength of light!
- Pigment **rhodopsin** dissociates—breaks down into smaller molecules
- "Bleaching reaction" is term used to describe reaction of rhodopsin
- Photoreceptor cells become permeable to ions resulting in depolarization and APs

Ion Flux Produces AP

AP passes through synapses from photoreceptor cells to

- bipolar cells and
- ganglion cells.

Axons of ganglion cells make up the optic nerve = CN II.

AP travels to the optic chiasma where the fibers partially cross to the opposite side of origin. Optic tract carries AP to the optic cortex in the occipital lobe of the cerebral cortex where vision is interpreted. The visual cortex is in the occipital area.

Go online and search for a graphic demonstrating the partial crossing of the optic nerves at the optic chiasma.

NOTES

Ear Physiology

Here are the foundational physiological concepts the novice A&P student needs to know about the ear.

✔ **CONCEPT!** The ear turns sounds waves in the air into sounds in our head and keeps us upright and stable!

The ear provides both hearing and equilibrium to the body. The middle and inner ear are involved with hearing while the vestibular apparatus of the inner ear is concerned with balance.

- Sound waves changed into action potentials
  - Sound waves = alternative areas of high and low pressure in the air
    - Molecules bunch up and then spread out in sound waves
  - Frequency determines the pitch of a sound
    - Waves closer together = higher pitch
    - Waves spread out = lower pitch
- Amplitude (height of wave) determines loudness of the sound
  - Greater amplitude = loud
  - Smaller amplitude = soft

- Air waves move the ear drum or tympanic membrane
- Tympanic membrane moves the three bones in the middle ear
  - Malleus
  - Incus
  - Stapes

- Stapes moves the oval window, which creates waves in the fluid inside the cochlea
- **Cochlea** is the **organ of hearing**!
- Fluid movement inside the cochlea moves the hair cells in the cochlea
- Hair cells have a type of cilia (remember that term?) on their surface
- Movement of the cilia causes an AP to fire off
- AP spreads into the vestibulocochlear nerve—CN VIII
- AP of CN VIII travels to the auditory cortex in the temporal lobe of the cerebrum

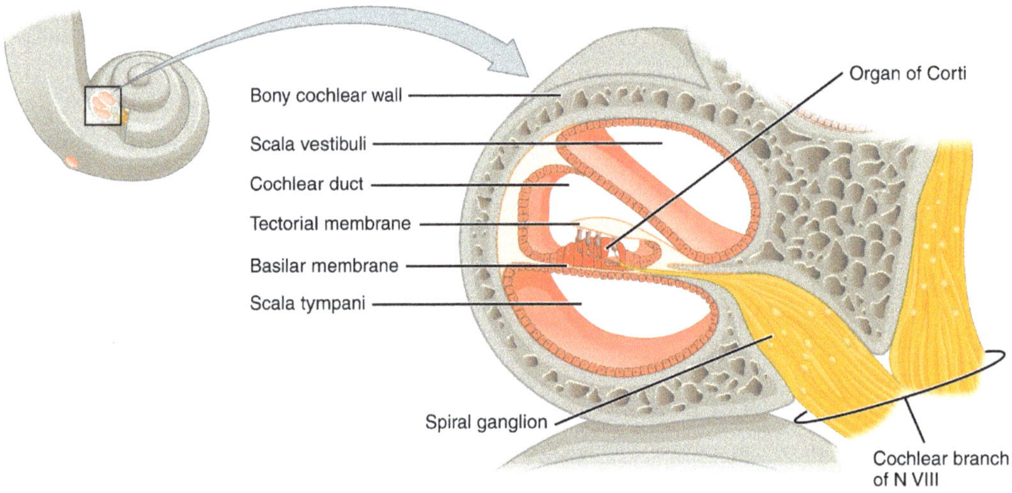

FIG. 12.25 Image of cochlea with cross section showing organ of Corti.

Keeping the Body Upright

Equilibrium and balance are two vitally important functions of the inner ear. The vestibular apparatus and semicircular canals keep the body upright and balanced.

- Motion is changed into APs
  - Inner ear (vestibular apparatus)—hollow organs
  - Three semicircular canals—tubes filled with fluid
- Otolith organs (cave-like shape) are filled with fluid and numerous tiny pebbles like grains of sand called **otoliths**
  - Utricle
  - Saccule

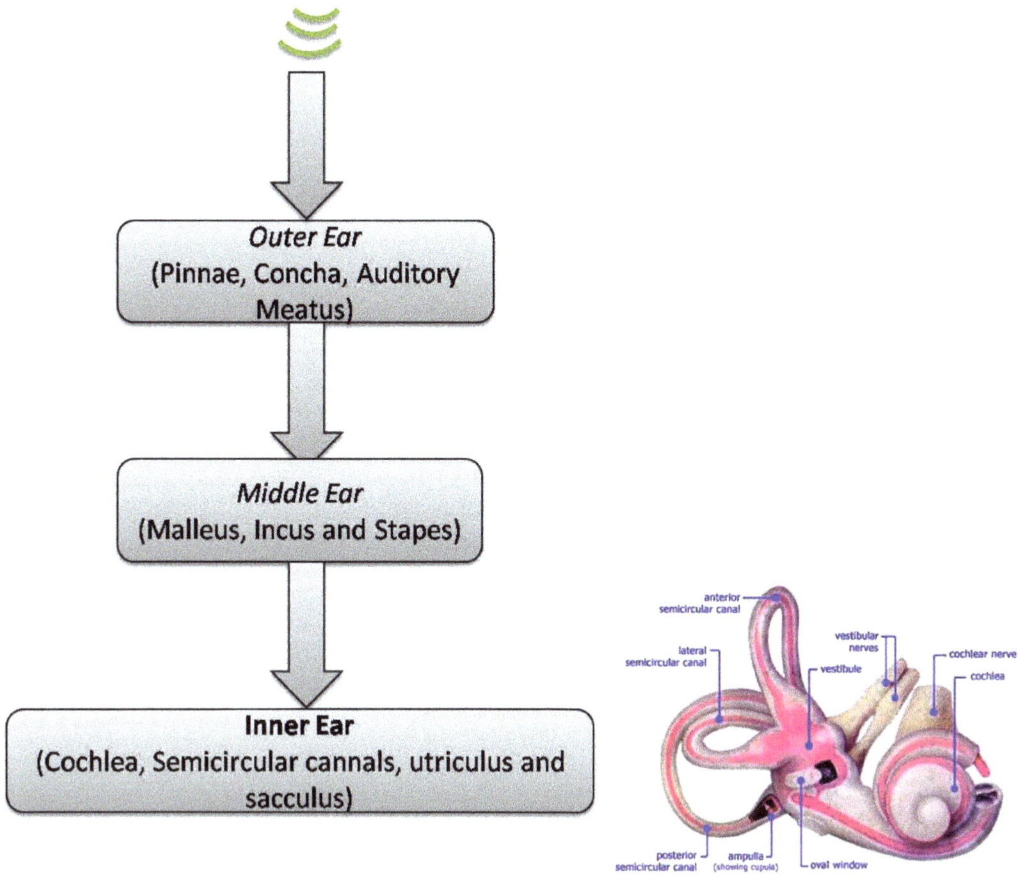

FIG. 12.26 Sound flowchart with graphic of inner ear organs.

These vestibular structures contain specialized hair cells much like the cochlea.

- Fluid in the semicircular canals, utricle, and saccule move when the head changes position
- Fluid movement stimulates the hair cells, causing APs
- APs travel through CN VIII to the brain stem where it is integrated with proprioceptor information from the cerebellum
- CNS processes sensory info and responds with appropriate motor output to maintain balance.
- Vertigo (dizziness)—loss of equilibrium—**can be devastating!**

Review Questions

- What is the body's communication system?
- How is the NS organized?
- What are the three meningeal layers surrounding the brain?
- What are the three parts of the neuron?
- What is a neuroglial cell?
- Conceptually, what do neuroglial cells do?

- The brachial plexus supplies innervation to what area of the body?
- Conceptually, how does the nervous system (NS) communicate?
- What makes up the central nervous system (CNS)?
- What makes up the peripheral nervous system (PNS)?
- What do each of the six neuroglial cells do?
- What are the basic physiological concepts of the eye and the ear?
- What are lamellar and tactile corpuscles
- Taste and smell involve what type of sensory receptors?
- Why are the eye and ear shaped the way they are?
- What is the basic function of the eye?
- What is the basic function of the ear?
- What cells produce myelin in the CNS? PNS?
- APs flow in what direction in the neuron?
- What is a synapse, and why is it an important structure?
- Which neurotransmitter has a unique structure?
- What ion causes depolarization in an AP?
- How does the eye turn light into images in the brain?
- What is the difference between structural and functional neurons?
- What are the functions of each type of neuroglial cell?
- What is an action potential (AP)?
- What ions diffuse during depolarization and repolarization in an AP?
- What is saltatory conduction?
- How does the ear turn sound waves into music in our brain?
- What do the otolith organs do?

Activities

- List the six types of neuroglial cells and how they support the NS.
- Grab those colored pencils or crayons and draw the cross section of the SC, showing white and gray matter as well as the dorsal and ventral horns and roots.
- Draw out and color the pathways of CN II, the optic chiasma, and optic tracts, showing which tracts cross contralaterally and which stay ipsilateral.
- Find an online site that shows how vertigo occurs. Watch it!
- Explain the importance of calcium ion in the NMJ, chemical synapse, and muscle contraction to someone who doesn't know A&P.
- Draw a graph of an AP showing RMP, threshold, depolarization, repolarization, and hyperpolarization.
- Teach someone else about the AP.

FIGURE CREDITS

Churn to Learn!

Teach to Learn!

Draw a Chemical Synapse on This Page

CHAPTER THIRTEEN

The Endocrine System

Glands, Hormones, and Homeostasis

TOES (TOPICS OF EMPHASIS)

- Exocrine versus endocrine glands
- Endocrine glands—pineal to testes
- Types of endocrine regulation or control
- Negative and positive feedback loops
- Endocrine hormones
- Blood glucose—insulin and glucagon
- Blood calcium—PTH and calcitonin

The NS is the body's primary communication system. The endocrine system is a secondary communication system. The NS communicates electrically, rapidly (almost instantaneously), but with very short duration of action. The endocrine system (ES) communicates chemically, relatively slowly, but with long duration of action. Figure 13.1 shows a simplified formula comparing the NS with the ES.

Nervous/Endocrine

COMPARISON

NS = C^5 = E/R/Shrt

ES = C^4 = C/Slow/Lng

FIG. 13.1 Comparison of NS with ES.

The ES is made up of eight glands: the pineal, pituitary, thyroid, parathyroid, pancreas, adrenal, ovaries, and testes. But maybe the most important part of the endocrine system is not a gland but an area of the brain, the hypothalamus. The anterior pituitary gland is referred to as the *master endocrine gland*, but the hypothalamus is the "master of the master." The hypothalamus secretes a variety of releasing and inhibiting factors (hormones), targeting the anterior pituitary and regulating the hormones it secretes. Figure 13.2 shows the endocrine glands and two endocrine areas, the hypothalamus and the thymus. When you're learning something like a list of the endocrine glands, always repeat them in the same sequence to help reinforce synapses and build solid learning neuronets.

SUPER Simple Summary of Endocrine Glands and Areas:

- Pineal gland—melatonin
- Hypothalamus—part of NS but major endocrine function; master of the master—secretes inhibiting and releasing hormones

- Pituitary—anterior (master endocrine gland) and posterior; ADH/Oxytocin; function as two separate glands
- Thyroid gland—thyroid hormone and calcitonin
- Parathyroid glands—four areas on the posterior aspect of thyroid gland; PTH
- Pancreas—dual exocrine (digestive enzymes) and endocrine (islet cells) function
- Adrenal glands—cortex and medulla; function as two separate glands
- Ovaries—dual endocrine/exocrine function; estrogen and progesterone/eggs
- Testes—dual endocrine/exocrine function; testosterone/sperm

It's important to compare endocrine glands with exocrine glands, like the salivary glands. Exocrine glands have ducts and secrete through a tube into another structure, usually another tube, like the intestines. Endocrine glands, on the other hand, do not have ducts. They secrete through the process of exocytosis into the interstitial space. Their secretions build up a concentration gradient and diffuse from the interstitial space into the blood stream (CVS) for circulation around the body where they target a variety of other glands or organs.

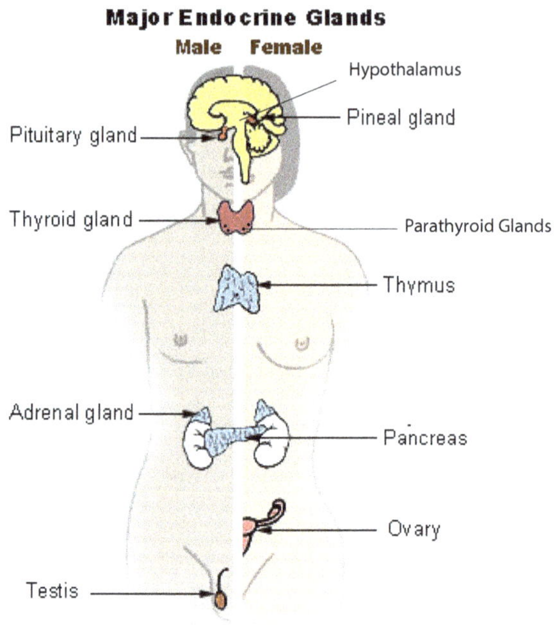

FIG. 13.2 Endocrine glands and areas.

Types of Endocrine Control

There are three types of **endocrine control** or regulation: humoral, hormonal, and nervous.

- **Humoral**—glands secrete in response to a change in the level of a nutrient or ion in the blood stream
    - Glandular tissue senses blood levels of substances like glucose or calcium
    - Respond to fluctuations until homeostasis is reached
    - Pancreatic islets and parathyroid gland
- **Hormonal**—glands release hormones when stimulated by other endocrine hormones: trophic hormones produced by hypothalamus and anterior pituitary
    - Utilize releasing hormones
    - TSH and ACTH are examples
- **Nervous**—glands release hormones in response to nervous stimulus
    - Pineal and retina
    - Posterior pituitary and hypothalamus
    - Adrenal medulla and sympathetic chain ganglion

Function of the endocrine system involves negative feedback loops (NFBLs). These loops were discussed earlier in chapter 10.

NFBLs always **reverse** the original stimulus. Positive feedback loops (**PFBLs**) always **amplify** the original stimulus. Several endocrine functions do involve PFBLs. Childbirth and lactation are endocrine functions that involve PFBLs. Blood clotting is also a form of positive feedback, but it is not an endocrine function.

Most endocrine hormones are regulated by negative feedback loops. Because it is essential at a novice level to understand concepts rather than memorize large amounts of isolated facts, we'll focus on the impact of endocrine hormones that control the homeostatic balance of blood glucose and blood calcium. More space will be devoted to blood glucose in this chapter because we covered the homeostatic balance of blood calcium in chapter 10 in relation to bone physiology.

✔ **CONCEPT!** At a novice level, understanding the concept of endocrine function and negative feedback loops is more important than struggling to memorize lists of hormones, where they are produced, and what they do.

NOTES

BRIDGE
to chapter 10, page 127 to review negative feedback loops.

Negative Feedback Loops and Glucose

Understanding endocrine control of blood glucose and calcium is an effective way to gain insight into the concept of endocrine function. Then it's easy to apply that concept in the future as you gain insight into other specific endocrine hormones.

Remember, all cells have a constant need for energy. This energy is created from the nutrients we ingest: C, P, L. Energy is produced by CR, as we discussed in chapter 6. Glucose is a breakdown of carbohydrates and is the primary source of energy production. Maintaining a homeostatic balance of blood glucose is essential for normal human function. This balance is accomplished by two antagonist endocrine hormones, insulin and glucagon.

BRIDGE
to Figure 2.1, the H_2O sheet, in chapter 2, page 13 as well as the discussion of CR in chapter 6 on page 57.

✔ **CONCEPT!** Insulin lowers blood sugar, whereas glucagon raises blood sugar levels!

Insulin is a polypeptide (protein) hormone produced by protein synthesis in the beta cells of the pancreatic islets (islets of Langerhans). Beta cells sense elevated blood sugar levels (humoral control) and respond by producing insulin. Insulin is packaged in vesicles (Golgi apparatus) and released through exocytosis into the interstitial space around the beta cells. Notice that insulin is released into the interstitial space and not directly into the blood stream. A small point, maybe, but an important concept to grasp.

✔ **CONCEPT!** Insulin is released into the interstitial space where it develops a concentration gradient and diffuses (simple diffusion) into the blood vessels of the pancreas.

Insulin is circulated throughout the body by the CVS. It lowers blood glucose levels by facilitating the transport of glucose from intravascular space, through interstitial space, into intracellular space. At this level, it's not important for you to understand the details of how insulin does that; just know insulin facilitates the transport of glucose from the blood vessels into the cells. Insulin is the key that unlocks the cell membrane so glucose can enter.

In the cells, glucose is either utilized to produce energy through CR or is converted to glycogen and stored for future use. Remember, glucose is a highly osmotically active solute, and osmotically active solutes "suck"! So, glucose accumulation inside cells would draw water into the cell (hypertonic solutions figuratively suck water) and cause the cells to swell and lyse (burst). Glycogen, however, is osmotically inert. So, glucose is converted into glycogen for storage.

✔ **CONCEPT!** Cells store glucose as glycogen because glycogen is osmotically inert!

The alpha cells in the pancreatic islets can sense (humoral control) when glucose levels are too low. The alpha cells then produce glucagon, another polypeptide chain (protein) manufactured by protein synthesis. Glucagon is then released into the interstitial space by exocytosis, just like insulin. It diffuses into the CVS and is transported primarily to the liver. In the liver, glucagon stimulates the conversion of glycogen back into glucose. The glucose then diffuses into the bloodstream, raising blood glucose levels.

Because the alpha and beta cells in the pancreatic islets can sense (humoral control) blood levels of glucose, they are able to increase or decrease their secretion of glucagon and insulin accordingly to maintain a homeostatic balance of blood sugar.

✔ **CONCEPT!** Insulin and glucagon maintain the homeostatic balance of blood glucose through humoral control and negative feedback loops!

BRIDGE
to chapter 10, page 127 to review calcium homeostasis.

The homeostatic control of blood calcium is another important example of endocrine function This was covered in chapter 10 when discussing bone physiology and the importance of maintaining a homeostatic balance of blood calcium. Go back to chapter 10 and review the process that maintains the homeostatic balance of blood calcium.

The endocrine system plays a major role in maintaining homeostasis in the body. Usually this involves negative feedback loops, but students need to be aware of the three areas where positive feedback loops contribute to normal body function.

- Uterine contractions during labor—oxytocin released from the posterior pituitary
- Lactation—prolactin from the anterior pituitary and oxytocin
- Blood clotting—cascade sequence resulting in clot formation

Figure 13.3 shows the positive feedback loops of laboring. Here, the original stimulus is amplified or increased until birth of the baby or expression of milk.

Types of Endocrine Hormones

Endocrine hormones are either protein or lipid based. An effective means to remember which are which is to memorize the six lipid or steroid-based hormones. Then, all the remaining endocrine hormones are proteins.

Lipid-Based or Steroid Hormones

- Adrenal cortical hormones
  - Aldosterone
  - Cortisol
  - Androgenic hormone (testosterone-like)
- Sex hormones
  - Estrogen
  - Progesterone
  - Testosterone

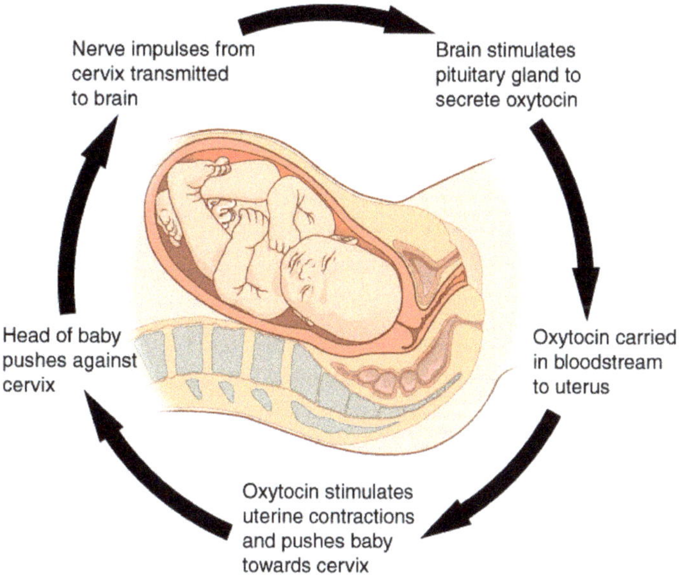

FIG. 13.3 Oxytocin and positive feedback during labor and delivery.

The steroid hormones move easily through the cell membranes of their target cells because the SPPLBL (cell membrane) contains lipids. The protein hormones do not directly enter the target cell but go through a transfer process to enter the target cell.

✔ **CONCEPT!** Steroid (lipid-based) hormones readily pass through the plasma membrane of target cells. Protein hormones go through a transfer process to enter target cells.

At an introductory level the two endocrine hormone pairs that are essential to know are insulin/glucagon and PTH/calcitonin. Obviously, there are many more endocrine hormones, and they all play important roles in human physiology. But at an introductory level it makes sense to focus on the two pairs we've discussed, allowing students to focus on grasping the **concept of endocrine control** rather than memorizing hundreds of facts relating to all the endocrine hormones.

Once you understand the concept of negative feedback loops and endocrine hormones, it's easier to grasp the specific details of other endocrine hormones that we do not focus on in *SS A&P*. For example, thyroid hormone produced by the thyroid gland targets all cells in the body. It regulates cellular metabolism and is under hormonal control.

Thyroid Hormone

- Protein-based (it's not one of the six steroid hormones)
- Hormonal control
  - TRH—Thyroid-releasing hormone from the hypothalamus
  - TSH—Thyroid-stimulating hormone from the anterior pituitary gland
- Negative feedback to hypothalamus and anterior pituitary gland as the blood level of circulating thyroid hormone fluctuates to maintain DC.

This is a good place for students to get online and watch the "Crash Course" segments with Hank Green that cover the endocrine system. Remember that to understand and gain insight, students need to look at any topic from a variety of perspectives. "Crash Course" can be very helpful!

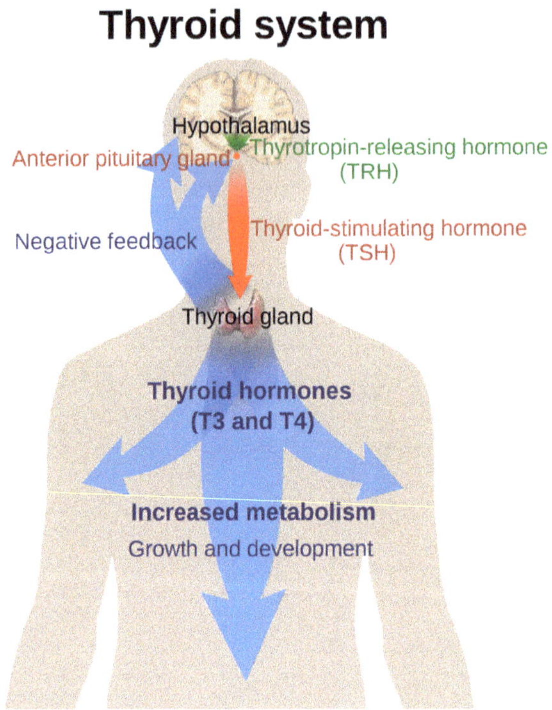

FIG. 13.4

NOTES

Review Questions

- What's the difference between an exocrine and an endocrine gland?
- What is a negative feedback loop?
- What is a positive feedback loop?
- What are three areas where positive feedback occurs?
- What antagonistic hormone pair balances blood glucose?
- What antagonistic hormone pair balances blood calcium?
- How does insulin lower blood glucose levels?
- Why is the homeostatic balance of blood glucose important?
- Why is the homeostatic balance of blood calcium important?

Activities

- Explain a negative feedback loop to a family member or a roommate.
- Describe the conceptual mechanism of action that allows insulin to lower levels of blood glucose.
- Describe in detail the mechanism of action allowing PTH to raise levels of blood calcium.
- Compare negative and positive feedback loops.

FIGURE CREDITS

Figure 13.2: Source: https://commons.wikimedia.org/wiki/File:Illu_endocrine_system.jpg.
Figure 13.3: Copyright © by OpenStax (CC BY 4.0) at https://commons.wikimedia.org/wiki/File:106_Pregnancy-Positive_Feedback.jpg.
Figure 13.4: Source: https://commons.wikimedia.org/wiki/File:Thyroid_system.svg.

Churn to Learn!

Teach to Learn!

UNIT IV

CIRCULATION, REGULATION AND REPRODUCTION

CHAPTER FOURTEEN

The Cardiovascular System

Structure and Function

TOES (TOPICS OF EMPHASIS)

- Relating the CVS to any transportation system
- Understanding the CVS flow chart
- Composition of blood
- Anatomy of the heart and great vessels
- Peripheral vessel concepts
- The primary function of the CVS
- CVS flowchart
- Blood—composition
- Heart—function and blood flow
- Differences between arteries and veins
- Relationship of lymphatic system to CVS
- Cardiac cycle (CC)
- Blood turbulence—S3, S4, murmurs and bruit

CVS = Cardiovascular System: The Body's Transportation System

The CVS is one of the most vital systems in the body. It has huge importance in clinical medicine and health care. Heart disease is now the major cause of premature morbidity (illness) and mortality (death) in America. Most, if not all, major hospitals in American have poured billions of dollars into the construction of freestanding heart hospitals. So, gaining basic conceptual insight into the structure and function of the CVS is essential for anyone pursuing a career in clinical medicine or any field related to health care. It is also a foundation for healthy living for all of us!

Remember those ten keys to learning in chapter 1? In the acronym REVV CCCAMM, the "A" stands for association. It is easier to learn new material if we can associate it, or relate it, to something we already know or understand. So, to help students grasp the foundational concept of the CVS, see Figure 14.1.

Any transportation system requires three things to function: vehicles, engines providing energy, and highways or tunnels where those vehicles can travel. This

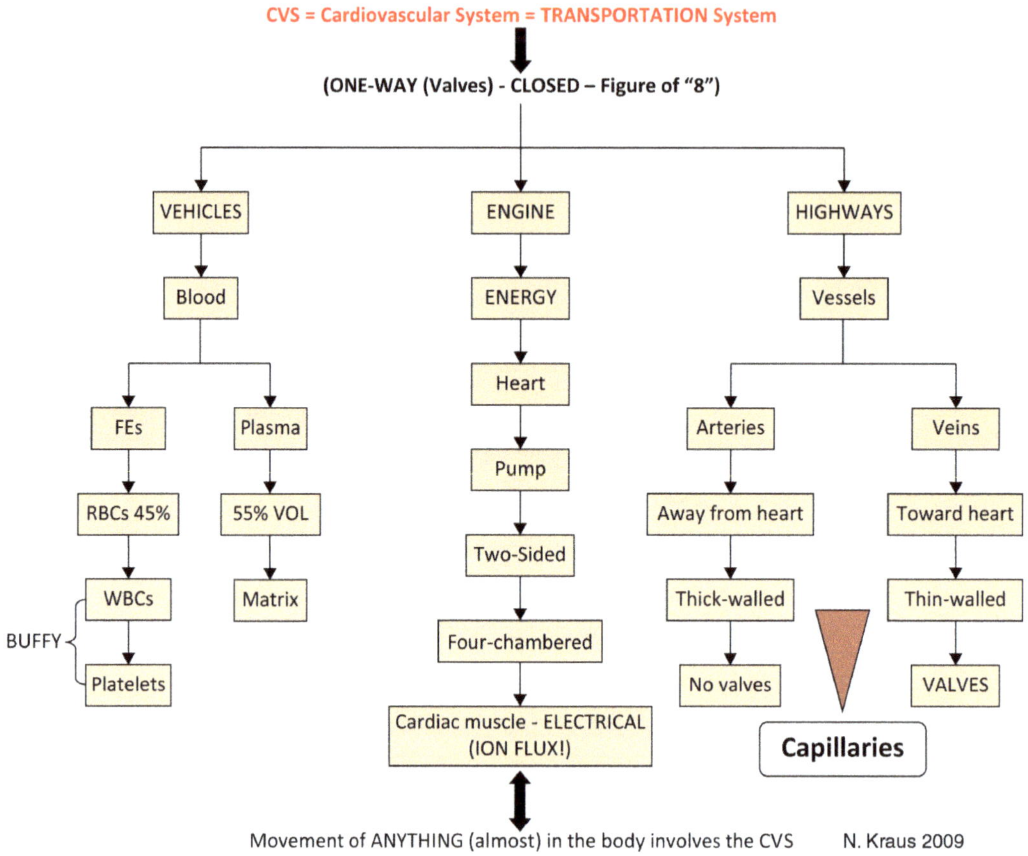

FIG. 14.1 Flowchart relating CVS to a transportation system.

is exactly what the human transportation system or CVS does. It has blood (the vehicles), the heart (an engine providing energy), and the vessels, arteries, and veins (where blood can travel throughout the body).

The CVS is a closed system; that is, the blood stays inside the heart and vessels. And it is a one-way system. Blood only flows in one direction. That one-way direction is maintained by the heart valves and valves in many veins. Finally, the CVS is a figure of eight system. That is, what goes out the right side of the heart comes back to the left side of the heart and what goes out the left side of the heart comes back to the right side of the heart.

Blood is a fluid connective tissue that has a matrix (plasma). Remember, all connective tissues have a matrix. The FEs, or formed elements, erythrocytes, leukocytes, and platelets, are suspended in the plasma.

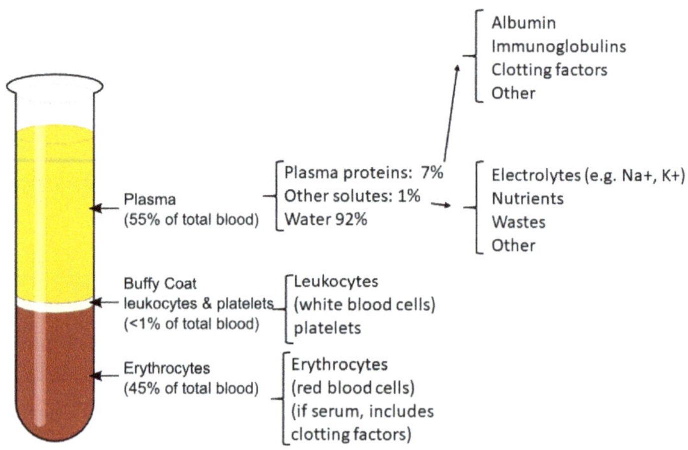

FIG. 14.2 Centrifuged whole blood showing components.

The RBCs (red blood cells) make up approximately 45 percent of whole blood volume because of the importance of what they do! They carry oxygen to the cells, where it is essential for the efficient production of cellular energy as discussed in Chapter 6.

The novice student needs to be able to recognize a whole blood smear and identify the FEs. Blood is the vehicle of the CVS. It transports or carries stuff, such as nutrients, oxygen, and wastes.

Go online and put "histology whole blood" into a browser. Spend some time getting comfortable with the appearance of whole blood and the blood cells (RBC, WBC, and platelets).

Blood travels in highways (or tunnels) called vessels. These are either arteries, that always carry blood away from the heart, or veins that always carry blood toward the heart. The arteries and veins are connected by capillary beds, which is where the real work of the CVS takes place. The capillaries are the site of the exchange of gases (oxygen and carbon dioxide), nutrients, and waste products produced during cellular respiration (production of energy). Without capillaries the CVS would be transporting stuff around the body, but the body wouldn't benefit because there wouldn't be any exchange of that stuff. It would be like trucks loaded with food driving around and around city streets but never stopping to unload at a grocery store. That's why the "real work" gets done at the level of the capillaries.

NOTES

The heart is the engine providing the energy to move blood throughout the body. It is two-sided (right and left) and has four chambers, two atria (receive blood), and two ventricles that eject or pump blood either to the lungs or the body. The right side of the heart is the pulmonary circuit, pumping blood to the lungs. The left side of the heart is the systemic circuit, pumping blood to the entire body (except the lungs).

Finally, the heart functions electrically. We measure that electrical energy or current flow with an ECG (EKG) or electrocardiogram. This shows the electrical activity of the heart but does *not* show the mechanical activity of the heart. The electrical energy causes the heart to contract and pump blood to the lungs and body. The ECG shows only the electrical impulses and not the mechanical contraction of the heart.

✔ **CONCEPT!** Anytime there is electrical current flow in the body it is caused by **ion flux**! An ion is an electrically charged particle and flux means movement, so ion flux means movement of electrically charged particles causing electric current.

The heart is structurally, functionally, and figuratively the "heart" of the CVS. It is primarily responsible for the movement of blood around the body! And although all the components of the CVS are essential for normal function, an argument can be made that the heart is the most essential organ in the CVS! Certainly, without the heart, the body doesn't function at all!

Essential Heart Structures to Know

It is helpful for students to utilize a comprehensive text or online source in conjunction with the guidance of *SS A&P*. Here are the structures you need to know about the anatomy of the heart.

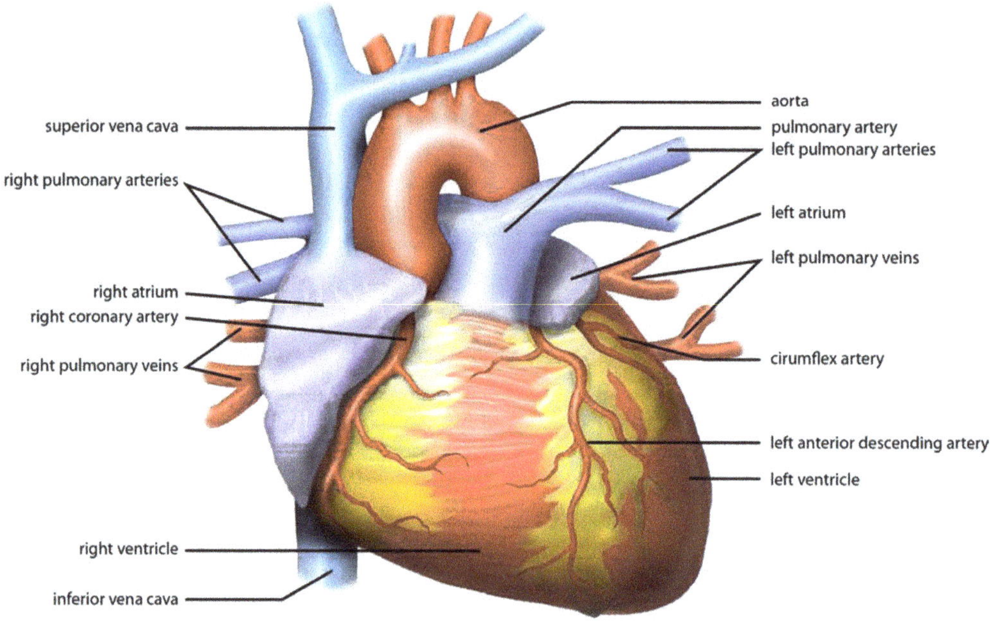

FIG. 14.3 Anterior view of the heart.

Anterior heart:

- Superior and inferior vena cava
- Pulmonary trunk
- Left and right pulmonary arteries
- Left and right pulmonary veins
- Right coronary artery
- Right auricle
- Right atrium
- Aorta
    - Ascending—two branches
        - Right coronary artery
        - Left coronary artery
            - Circumflex artery
        - Anterior interventricular or left anterior descending artery
    - Arch—three branches
        - Brachiocephalic trunk

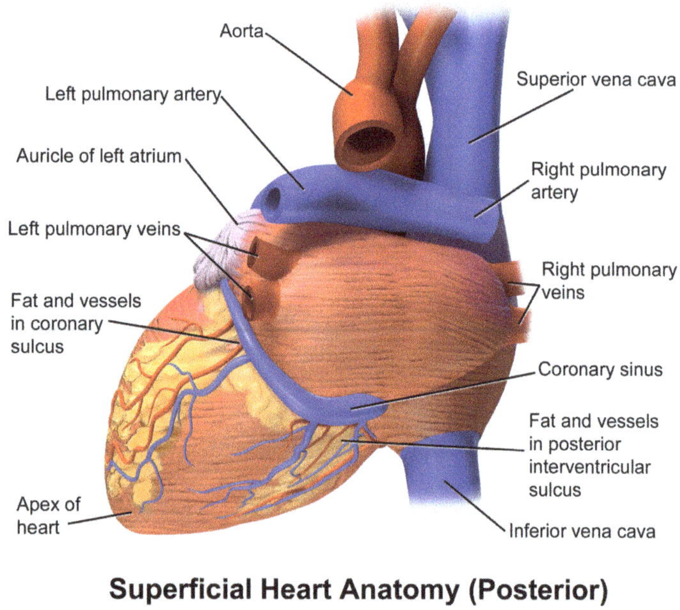

FIG. 14.4 Posterior view of the heart.

- Left common carotid artery
- Left subclavian artery

- Ligamentum arteriosum—remnant of the ductus arteriosum in the fetus

The posterior view has most of the same structures as the anterior view but adds the coronary sinus.

Students need to know all the labeled structures and chambers of the internal anatomy of the heart:

- Aorta
- Superior vena cava
- Right pulmonary artery
- Pulmonary trunk
- Right pulmonary veins
- Right atrium
- Fossa ovalis
- Tricuspid valve
- Right ventricle
- Chordae tendineae
- Inferior vena cava
- Left pulmonary artery
- Left atrium
- Left pulmonary veins
- Mitral (bicuspid) valve
- Aortic valve
- Pulmonary valve
- Left ventricle
- Papillary muscles
- Interventricular septum
- Epicardium
- Myocardium
- Endocardium

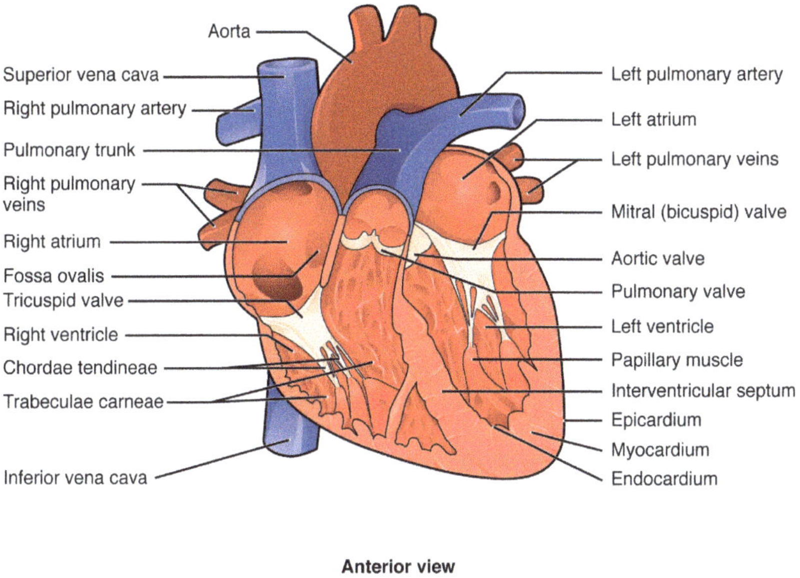

FIG. 14.5 Interior view of the heart.

Pericardial Sac

All internal organs or viscera are covered by an epithelial serous membrane (epithelium lines and covers, remember?). This single serous membrane lines the closed body cavity, creating a visceral and a partial sheet of epithelium. The heart, however, has a unique pericardial sac surrounding it, as shown in Figure 14.6.

The pericardial sac has two distinct layers (rule of two!), a tough outer fibrous layer and a thin inner serous layer. The fibrous layer consists of dense irregular connective tissue,

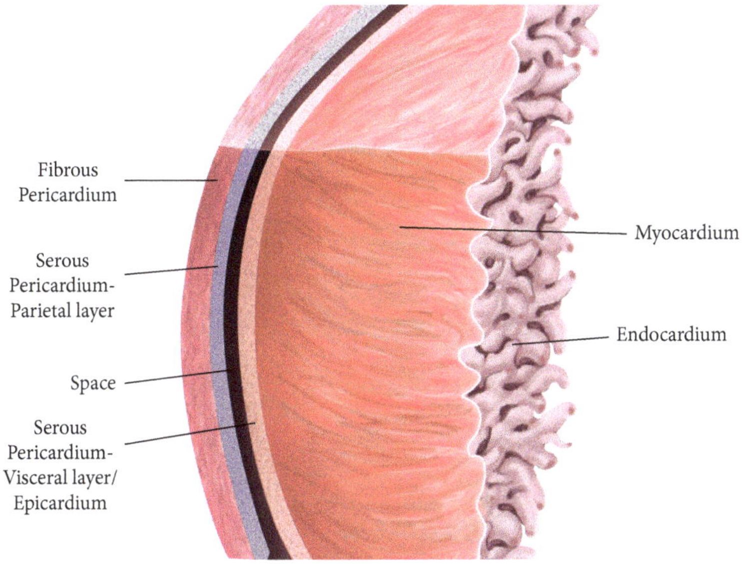

FIG. 14.6 Pericardial sac.

which will not stretch. It prevents the heart from overfilling but can also have serious clinical consequences (cardiac tamponade).

✔ **CONCEPT** Rule of two: the pericarium has two layers, serous and fibrous

The serous pericardium also has two layers (rule of two again!): the outer or parietal layer, which lines the inside of the fibrous pericardium, and an inner visceral layer on the outside of the myocardium (also called the epicardium).

The space between the visceral and parietal pericardium is the pericardial space. It contains a small amount of serous fluid providing lubrication, which allows the heart to freely move as it contracts and relaxes.

Arteries and Veins

Arteries always carry blood away from the heart (but not always blood high in oxygen content!).

Arteries and veins have different structures because of what they do!

- Arteries have thick walls made mostly of visceral muscle
  - Able to divert blood to specific areas when needed
  - More circular than veins
  - Vascular sphincters (visceral muscle) shunt blood
  - Able to withstand higher blood pressure
- Veins have thin walls and less visceral muscle
  - Have lower pressure than arteries
  - Contain valves to help maintain one-way blood flow
  - Larger but less structured than arteries
- Both arteries and veins have three tunica (layers)
  - Tunica externa
  - Tunica media
  - Tunica intima

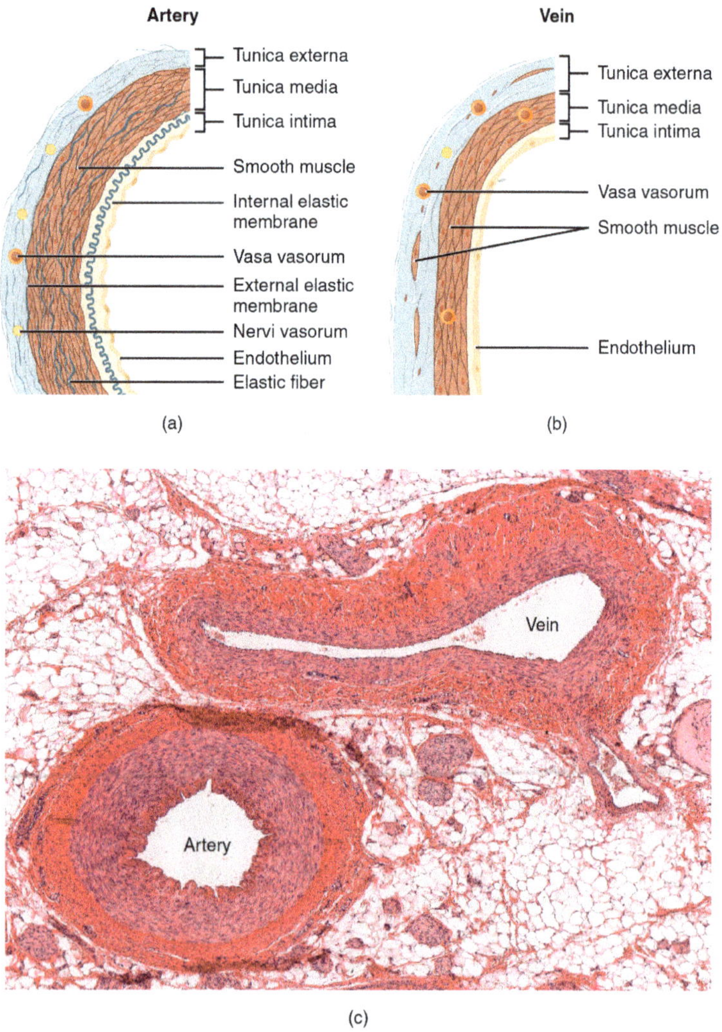

FIG. 14.7 Comparison of an artery and a vein.

Figure 14.7 demonstrates the anatomy of arteries and veins. The novice A&P student only needs to focus on the main arteries of the body.

These are demonstrated in Figure 14.8.

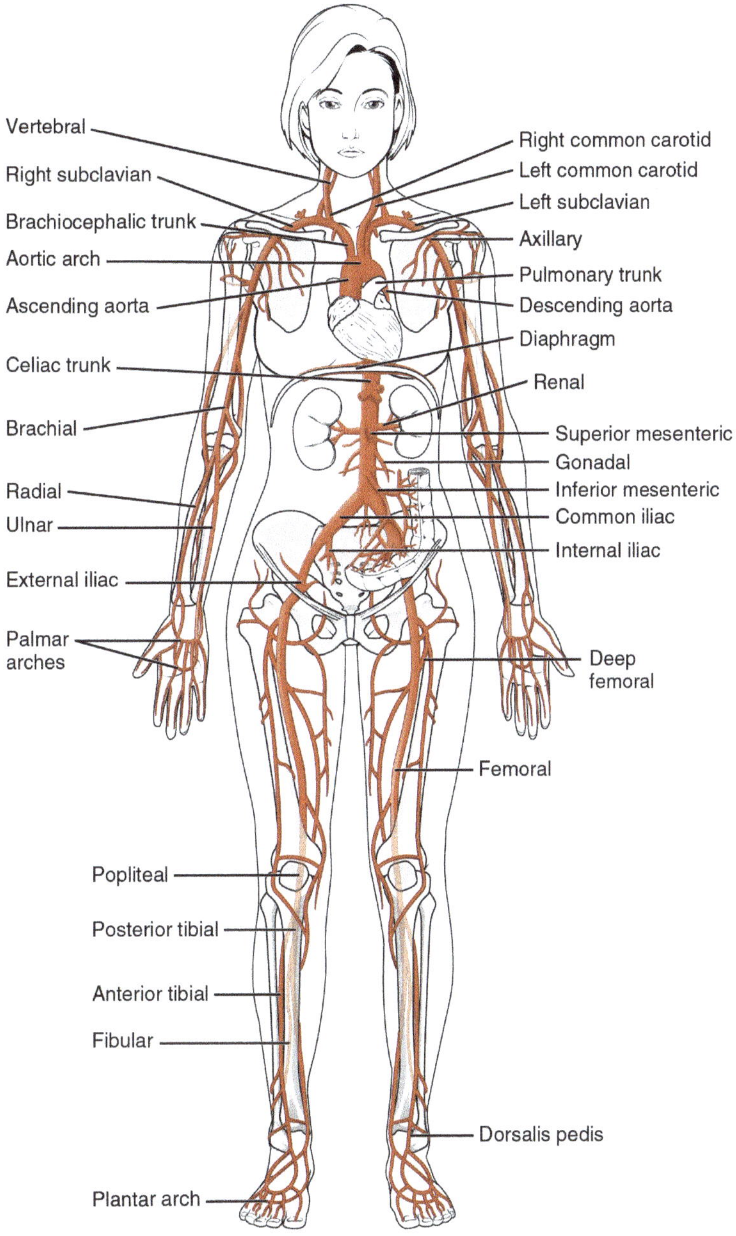

FIG. 14.8 Major arteries of the body.

✔ **CONCEPT!** Once you've mastered the main arteries of the body, it's easy to learn the main veins because most deep arteries (those not near the surface of the body) have a vein running parallel to the artery. For example, there is a brachial artery and a brachial vein; a femoral artery and a femoral vein; a common carotid artery and an internal jugular vein.

There are several superficial veins the introductory student needs to know:

- **Cephalic vein**—anterior aspect of the arm over the lateral aspect of the biceps brachii
- **Median cubital vein**—anterior aspect of the elbow; frequent blood draw site
- **Great saphenous vein**—longest vein in the body
  - Runs from the great toe to the groin where it empties into the femoral vein
  - Frequently utilized as a detour vessel during coronary bypass surgery

FIG. 14.9 Demonstration of cephalic vein in left brachial area.

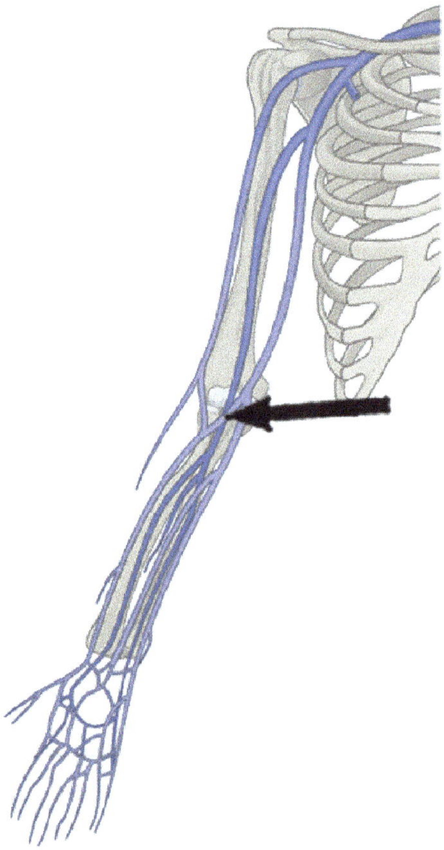

FIG. 14.10 Median cubital vein.

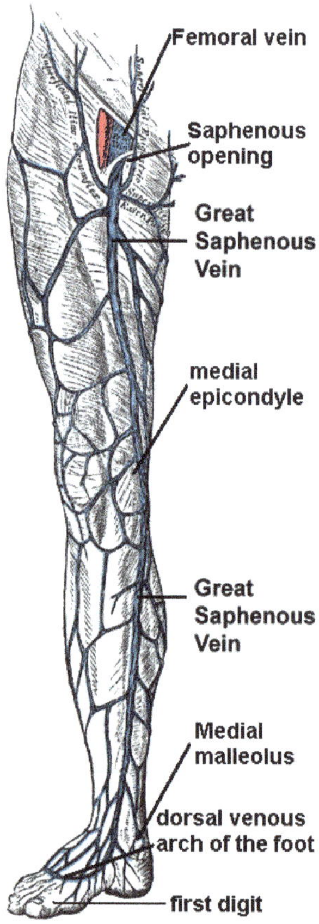

FIG. 14.11 Lower extremity showing great saphenous vein.

Deep Vein Exceptions

As happens with most general concepts, there are some exceptions regarding specific deep veins that are important for the novice A&P student to know.

- **Hepatic portal vein**—a large vein found in the abdominal cavity. It is formed by three veins, the superior and inferior mesenteric veins, and the splenic vein. Notice in Figure 14.12 the smaller veins that drain into these three larger veins (e.g., the pancreatic, colic, and sigmoid veins, etc.). An introductory student does not need to identify these smaller veins. However, students should be able to deduce the area drained by each vein. This is a great example of the concept of common sense in A&P. The names tell you exactly what the vein drains. For example, the pancreatic vein drains the pancreas and the sigmoid vein drains the sigmoid colon!

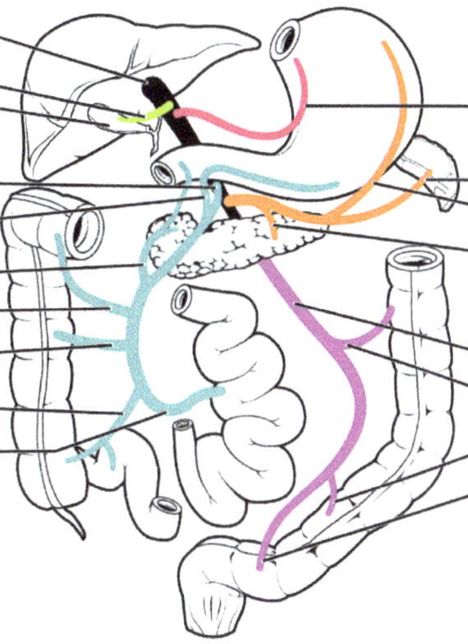

FIG. 14.12 Hepatic vein and contributing tributaries.

✔ **CONCEPT** Common sense: the names of many vessels reveal the organ supplied or drained

Portal venous systems have a dual (two) capillary bed. In the hepatic portal system, the first capillary bed is formed by branches of the gastric, splenic, superior, and inferior mesenteric arteries. This first capillary bed surrounds the intestines and is responsible for the absorption of nutrients from the intestine. This capillary bed then drains into the branches of the hepatic portal vein, which drains into the liver. Blood in the hepatic portal system is high in carbon dioxide, low in oxygen content, and high in nutrient content. The second capillary bed in the hepatic portal system is the sinusoids inside the liver. Blood then drains out of the liver through the hepatic vein into the inferior vena cava.

- **Azygos vein**—drains the posterior aspect of the thoracic cage emptying into the superior vena cava. The hemiazygos and accessory hemiazygos veins drain into the azygos forming the azygos system. The azygos vein forms an alternative route (a detour) for blood from the inferior vena cava to the superior vena cava, which can have clinical significance. Go online and find a graphic showing the azygos venous system.

Lymphatic System

The lymphatic system is part of the immune system. Its primary function is to provide defense for the body. We'll look at the immune system in detail in chapter 15.

However, the lymphatic system has a secondary function of absorption of interstitial (between the cells) fluid, returning that fluid to the CVS. It does this primarily through the thoracic duct that drains interstitial fluid from most of the body into the left subclavian vein.

BRIDGE to Chapter 15, Lymphatic System

The right lymphatic duct drains interstitial fluid from the right side of the head and neck, the right half of the thoracic cage, and the right upper extremity into the right subclavian vein.

In addition, blind lymphatic capillaries (lacteals) in the villi of the small intestine function to absorb lipid molecules from the gut. These fat molecules are too large to be easily absorbed into the intestinal capillary beds. These absorbed fats are added to the resorbed interstitial fluid to form lymph, one of the types of fluid connective tissue.

BRIDGE to Chapter 8, Histology, Connective Tissue on page 82

Figure 14.13 shows the lymphatic vessels throughout the body. Make sure you know the location and drainage area of both the thoracic duct and the right lymphatic duct.

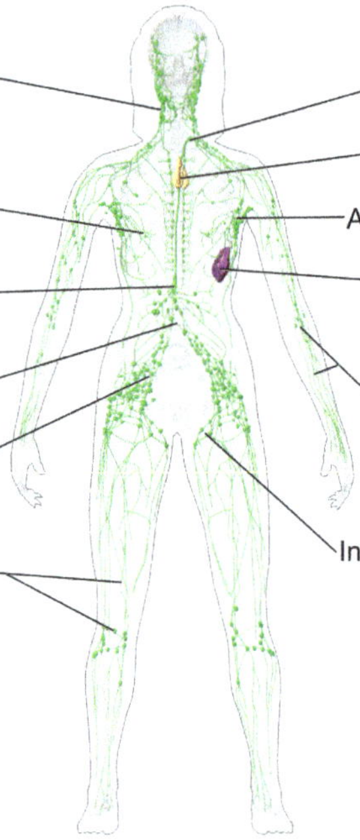

FIG. 14.13 Vessels of the lymphatic system.

NOTES

Cardiovascular System Physiology

Remember, the CVS and the heart are structured the way they are because of what they do! As the CVS flowchart (Fig. 14.1) states, the heart is the "engine" of the body's transportation system. It is the primary source of energy moving the blood (vehicles) through the highways or vessels. The heart is structured the way it is to facilitate the efficient circulation of blood to every organ in all parts of the body.

✔ **CONCEPT!** The heart is shaped the way it is because of what it does!

Blood Flow through the Heart

As the CVS flowchart (Figure 14.1) shows, the heart is two-sided. The right side or the pulmonary circuit, receives deoxygenated blood from the body. It then pumps blood out to the air sacs of the lungs where carbon dioxide and oxygen are exchanged. The oxygenated blood then flows back to the left side of the heart.

The left side of the heart is the systemic circuit receiving oxygenated blood from the lungs and pumping it to the body. Anatomically the lungs are located right next to the heart, so the right ventricle doesn't have to pump with high pressure. But the left side of the heart pumps to the entire body over greater distances and so has a much higher pressure, about ten times higher than the right side! That's why the myocardium of the left ventricle is much thicker than the wall of the right ventricle. It works much harder.

✔ **CONCEPT!** Organs are structured as they are because of what they do. The left ventricle has much more muscle (thicker wall) because of what it does, which is pump blood to the entire body.

It is essential for the novice student to learn how to trace the flow of blood through the heart, to the lungs, back to the heart, out to the body, and back to the heart. Let's start with an RBC, "Robbie red cell" located in the right atrium and follow him on his round-trip journey.

R. atrium— R. AV valve (tricuspid)—R. ventricle—pulmonary semi-lunar valve—pulmonary trunk (low oxygen content)—pulmonary arteries (L. and R.)—lungs (gas exchange O_2 and CO_2)—pulmonary veins (high oxygen content)—L atrium—L. AV valve (bicuspid or mitral)—L. ventricle—aortic semi-lunar valve—ascending aorta (L. R. coronary arteries)—arch of the aorta (brachiocephalic trunk, L. common carotid, L. subclavian artery)—descending aorta—body (exchange of gases, nutrients, and wastes)—superior vena cava (SVC)—inferior vena cava (IVC)—coronary sinus (draining myocarium)—R atrium.

That's quite a trip for Robbie, but he does it thousands of times a day, and you need to be able to trace Robbie's trip itinerary confidently.

Coronary Blood Flow

It's important for the introductory student to know that the coronary arteries fill with blood during diastole and not systole. This is because the openings of both the left and right coronary arteries are located within the cusps of the aortic semi-lunar valve. So, during systole when the aortic valve is open, the openings of the coronary arteries are blocked by the cusps of the aortic valve. But even if they weren't blocked, blood would not be able to perfuse the ventricular myocardium because the heart muscle is contracted, compressing the blood vessels in the heart muscle.

Go online and search for animations demonstrating the location and filling of the coronary arteries to deepen your understanding of this unique characteristic of blood flow to the heart.

NOTES

Arteries and Veins

The "highways" of the CVS are the arteries and veins. Arteries always carry blood away from the heart and veins always carry blood toward the heart. However, arteries don't always carry blood high in oxygen content and veins don't always carry blood low in oxygen content. Think about Robbie's trip. When he's leaving the right side of the heart (pulmonary circuit) he's low in oxygen, having given up most of his oxygen to the tissues of the body for cellular respiration. Robbie goes to the lungs and gets rid of his load of CO_2 (waste from CR) and takes on a load of O_2. Now this oxygen-rich blood flows back to the heart through the pulmonary veins.

Blood that is high in oxygen content is scarlet in color, and that's why most arteries on anatomical models are a reddish color. Blood that's low in oxygen content appears to be a darker red or purple color, and that's why most veins on anatomical models are bluish in color. The pulmonary trunk and arteries will be colored blue on anatomical models and the pulmonary veins scarlet, because of the unusual reversal of typical oxygen content in the pulmonary arteries and veins. Make sure you notice this characteristic of heart models in the A&P laboratory or on graphics of the heart and vessels (Fig. 14.3, 14.4 and 14.5).

✔ **CONCEPT!** Arteries carry blood high in oxygen content away from the heart. Veins carry blood low in oxygen content toward the heart. But not always! **Know the exceptions!**

Cardiac Cycle

The cardiac cycle (CC) is an important learning tool that initially may appear overwhelming but with some patience and study becomes helpful for the novice student. The cardiac cycle shows all the events occurring in the heart within one heartbeat. See Figure 14.14.

The CC consists of six graphs superimposed on each other. Three of those graphs show pressure, one shows ventricular volume, one shows the electrical activity in the heart (EKG), and one shows an audiogram or phonocardiogram of the heart sounds. The CC refers to the left side of the heart, which is the high-pressure systemic division pumping blood to the entire body except the lungs.

The CC has two main phases, **systole** and **diastole**. Systole is when the heart muscle is contracting and ejecting blood. Diastole is when the heart is relaxed and filling with blood. Both the atria and ventricles have systolic and diastolic phases, but ventricular systole is more important than atrial systole. Notice diastole, ventricular filling, is about twice the length of time as systole during ventricular contraction.

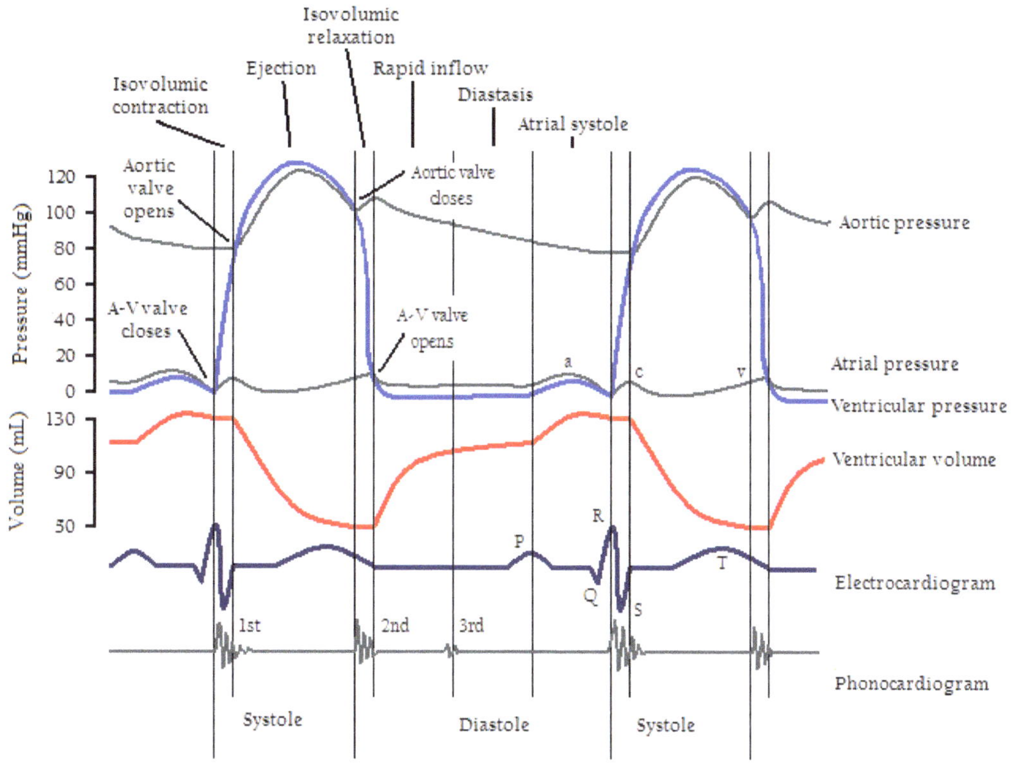

FIG. 14.14 Cardiac cycle.

The three pressure graphs show pressure in millimeters of mercury (mmHg) for the following cardiac chambers or structures:

- Aorta (AP)
- Left ventricle (LVP)
- Left atrium (LAP)

Strive to figure out the meaning of the small letters "a," "c," and "v" right above the atrial pressure curve in Figure 14.11.

The volume graph shows how the left ventricular volume changes as blood is ejected from the heart and as the ventricle fills up with blood during diastole.

The **EKG** shows the **electrical activity** of the heart. It does *not* demonstrate any mechanical or muscle contraction activity of the heart. Yes, the electrical depolarization of the heart initiates contraction, but the EKG only shows electrical activity and not contraction.

✔ **CONCEPT!** The EKG shows only the electrical activity of the heart.

The first wave or bump in the EKG is the "P" wave and shows atrial depolarization. This will cause the atria to contract, but remember that the EKG only shows cardiac depolarization (electrical activity) and not atrial contraction. The next wave complex is the "QRS" wave, showing ventricular depolarization. Finally, the last bump is the "T" wave, demonstrating ventricular repolarization.

At this stage of learning A&P, strive to gain an understanding of the concept of the EKG. Don't worry about reading an EKG; just grasp the concept and what the basic waves indicate. When, and if, you need to learn how to read EKGs, a great learning tool for that is ***Rapid Interpretation of EKGs*** by Dale Dubin, MD. It is a simple and straightforward programmed learning text that will have you reading EKGs in one weekend of study! It's simply the single best medical text this author has ever seen! There are also online sources which are beneficial, including a site for *Rapid Interpretation of EKGs*!

The sixth and final graph of the CC shows the heart sounds: first, second, third, and fourth. The first two heart sounds (S1 and S2) are caused by the closing of the heart valves. S1 occurs when the pressure in the ventricles becomes greater than the pressure in the atria at the beginning of systole. *Fluids, including blood, flow from higher to lower pressure,* so blood starts to flow backward into the atria. That's wrong-way flow! Remember, as the CVS flowchart indicates, the CVS is a one-way system. That one-way flow is maintained primarily by the heart valves. So, the AV valves slam shut, preventing wrong-way flow, maintaining right-way flow, and creating the "lub" sound of S1.

Then, at the end of systole and the beginning of diastole, the pressure in the ventricle falls below the pressure in the aorta, so blood wants to flow back into the ventricle from the aorta. Once again, that's wrong-way flow! Now the aortic semi-lunar valve (why are they called "semi-lunar"?) slams shut to maintain right-way blood flow, causing the "dub" sound of S2.

NOTES

S1 and S2 are caused by valve closure. Any other sound (with a rare exception) in the heart, or anywhere in the CVS, is caused by blood **turbulence**! When fluids flow in a straight, laminar pattern they do not make noise. But when fluids tumble, they make noise. In the normal CVS, blood flow is laminar and so soundless. When there are abnormalities or a large volume of blood flow in the heart or blood vessels, blood will tumble and thus make noise.

✔ **CONCEPT!** S1 and S2 are caused by valve closure. CVS sounds other than S1 and S2 are caused by blood turbulence!

S3 and S4 are examples of sounds caused by blood turbulence. In a fully compliant and vigorous heart like you'd find in most children and athletes, early diastolic filling is so rapid it creates some blood turbulence as blood rushes passively from the atria into the ventricles during early diastole. This causes a soft sound called the third heart sound or S3: "lub, dub, bup ... lub, dub, bup ... lub, dub, bup." It's called an S3 gallop because it sounds like a horse galloping, once again demonstrating the fact A&P are logical and filled with common sense.

✔ **CONCEPT** Common Sense: S1 and S2 combined with an S3 make a sound like a horse galloping, so the S3 is called a "gallop"

S3 can many times be heard in a normal heart, but the S4 is most likely a pathological sound caused by a non-compliant ventricular wall in late diastole. For example, a heart attack (MI) would cause death of some ventricular muscle, creating scar tissue. Normal heart muscle is compliant, able to easily stretch, contract, and relax. Scar tissue, on the other hand, is non-compliant. It's stiff. So, when the atria contract (atrial systole) during late diastole, packing more blood into the ventricles, the scar tissue doesn't stretch as it normally would, causing some blood turbulence and an S4 gallop.

Again, this is another great opportunity to go online and research S3 and S4 gallops. You can find sites that will play an audio of these gallops, helping your understanding. Listening will also help you be more comfortable when (and if) you are in a clinical setting auscultating a patient's heart or when discussing heart sounds with preceptors and colleagues.

A final example of blood turbulence is Korotkoff sounds. These are the sounds heard when taking a patient's blood pressure. When taking BP in the upper extremity, the cuff is inflated above the systolic pressure of the left ventricle, completely closing off the flow of blood in the brachial artery. As the cuff pressure is slowly released it will reach and then drop below the systolic ventricular pressure. This will allow a small squirt of blood to push through the brachial artery. This causes blood turbulence and a sound, which indicates systolic blood pressure. As the cuff pressure decreases and passes the point of diastolic pressure, the blood returns to laminar (straight) flow and the sounds disappear. These Korotkoff (Russian who discovered them) sounds indicating BP are caused by blood turbulence.

Blood flow through a normal common carotid artery is laminar. When a stethoscope is placed over a normal, healthy artery (like the common carotid), no sound will be heard. However, if a "whoosh ... whoosh ... whoosh" sound is heard, this is a **carotid bruit**. The sound is caused by blood turbulence. Something (a narrowing or an aneurism) is causing the blood to tumble and make a sound, which is heard on auscultation.

✔ **CONCEPT** Sounds other than S1 and S2 are caused by blood turbulence

---

Any sound other than S1 and S2 anywhere in the CVS (heart, artery, vein) is caused by blood turbulence (with a rare exception). Now, with good A&P understanding it will be easy to figure out what specific lesion might cause turbulence! With conceptual understanding you'll be able to figure stuff out!

It is also important to understand the concepts of stroke volume and cardiac output. Stoke volume (SV) is the amount of blood pumped out of the heart by the ventricle in a single heartbeat. The SV can also indicate the ejection fraction (EF), which is the percentage of blood pumped by the ventricle in a single heartbeat. Remember, the chambers of the heart are always completely filled with blood. The volumes change as the chambers contract and relax, pumping and receiving blood. So, if the EDV (end diastolic volume) is 100 cc, and the SV is 70 cc, the EF would be 70 percent.

Cardiac output (CO) is the total amount of blood pumped out of the heart (ventricles eject blood, atria receive blood) in one minute. So, a patient with a stroke volume (SV) of 70cc and a heart rate (HR) of 70 would have a CO of 4.9 liters, or 4,900 cc or 4,900 ml. Remember, if you're going to work anywhere in a clinical setting you must understand the metric system.

CO can be expressed as a mathematical formula: $CO = SV \times HR$.

NOTES

Review Questions

- In a single word, what's the primary function of the CVS?
- How does the CVS compare to transportation systems in our communities?
- Blood is which of the primary tissues?
- How does the lymphatic system augment the CVS?
- Why is it important to understand the CC?
- What causes all CVS sounds other than S1 and S2?
- Why is the left ventricle of the heart thicker than other heart chambers?
- What is the significance of the CVS flowchart?
- Lymph capillaries drain fluid from what fluid space?
- Why is the concept of blood turbulence important?
- The right side of the heart pumps blood to what organs?
- What artery is known as the "widow maker"?
- What structures maintain the one-way flow of blood in the CVS?

Hopefully, you've been striving to figure out the a, c, v in the atrial pressure curve of the CC, here's the explanation. The "a" corresponds to atrial contraction, wave "c" corresponds to an increase in pressure from the closed mitral valve bulging into the atrium during ventricular systole, and the wave "v" corresponds to passive atrial filling, causing a slight increase in pressure.

Activities and Coloring

- With colored pencils/crayons draw the circulation of blood flow through the heart, to the lungs, back to the heart, out to the body, and back to the heart. Label your drawing. Do this conceptually, simply, and graphically. It doesn't have to be anatomically correct. It can be drawn as a box divided into four squares representing the four chambers of the heart.
- Explain blood turbulence to a student from another A&P section or one of your classmates, your roommate, a friend, or a family member.
- Explain the CVS flowchart to someone who has never seen it before. This could be a family member (maybe a parent) or a sibling or roommate. It's even better if that roommate is not a science major and doesn't know any A&P. You'll learn a ton teaching them.

✔ **CONCEPT!** If you really want to understand something, struggle teaching it to someone else.

FIGURE CREDITS

Figure 14.6: TeachMeSeries Ltd, http://teachmeanatomy.info/thorax/organs/heart/pericardium.

Figure 14.7: Copyright © OpenStax College (CC BY 3.0) at https://commons.wikimedia.org/wiki/File:2102_Comparison_of_Artery_and_Vein.jpg.

Figure 14.8: Copyright © OpenStax College (CC BY 3.0) at https://commons.wikimedia.org/wiki/File:2120_Major_Systemic_Artery.jpg.

Figure 14.9: Copyright © 2017 Depositphotos/AlexFedorenko.

Figure 14.10: OpenStax College, http://anatomyzone.com/wp-content/uploads/2015/04/medial-cubital-vein.png. Copyright © by Rice University. Reprinted with permission.

Figure 14.11: Source: https://commons.wikimedia.org/wiki/File:Great_saphenous_vein.png

Figure 14.12: Copyright © by OpenStax College (CC BY 3.0) at https://commons.wikimedia.org/wiki/File:2138_Hepatic_Portal_Vein_System.jpg.

Figure 14.13: Copyright © by BruceBlaus (CC BY 3.0) at https://commons.wikimedia.org/wiki/File:Blausen_0623_LymphaticSystem_Female.png.

Figure 14.14: Copyright © DanielChangMD (CC BY-SA 2.5) at https://commons.wikimedia.org/wiki/Category:Wiggers_diagram#/media/File:Wiggers_Diagram.png.

Figure 14.15: Copyright © by User:Wapcaplet (CC BY-SA 3.0) at https://commons.wikimedia.org/wiki/File:Heart_numlabels.svg.

Churn to Learn!

Teach to Learn!

Fill in the Indicated Labels on this Diagram of the Heart

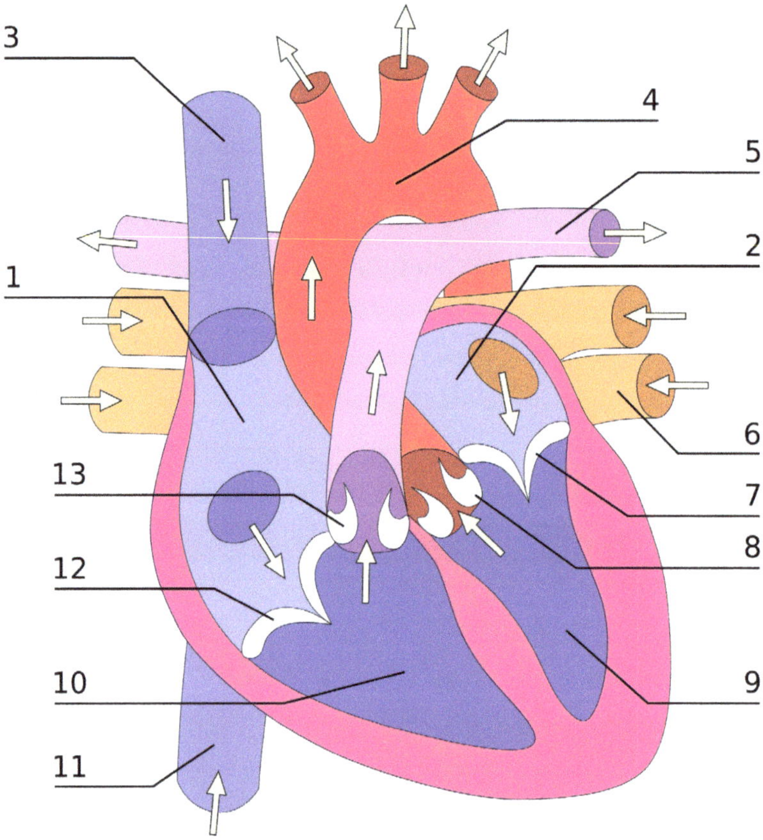

FIG. 14.15 Labeling Structures of the Heart.

CHAPTER FIFTEEN

The Lymphatic and Immune Systems

Protection or Defense

TOES (TOPICS OF EMPHASIS)

- Overall immunity or body defense
- Innate immunity
- Acquired or adaptive immunity
- Antigens and antibodies
- Active and passive immunity

Our bodies are constantly exposed to a variety of potentially harmful agents in the environment. The immune system provides protection for the body. It is the body's defensive system.

✔ **CONCEPT!** **Immunity** is all about **defense**!

There are two types of body defense the novice student needs to know, innate immunity and adaptive or acquired immunity.

Innate Immunity

Innate immunity provides protection for a wide variety of nonspecific hazards and pathogens. We are born with innate immunity, which forms barriers to prevent entry of potentially harmful substances and responds to a wide variety of harmful agents before or after they've entered the body. Innate immunity occurs immediately or in a short period of time.

There are two types of innate immunity (rule of two), barrier and non-barrier.

✔ **CONCEPT** Rule of two: There are two types of innate immunity

Examples of nonspecific barriers that provide innate immunity include the following:

- Skin
- Mucous membranes

- HCL (hydrochloric acid) in the stomach
- Lacrimal fluid—tears

Examples of non-barrier innate immunity include the following:

- WBCs—neutrophils, basophils, mast cells
- NK, or natural killer cells—detect and destroy unhealthy cells
    - Cells infected with bacteria
    - Tumor or cancer cells
    - NK cells are formed in bone marrow
    - NK cells circulate in the blood
    - NK cells accumulate in spleen, lymph nodes, tonsils
- Interferons—protect against spread of viral infections
- Complement system—thirty types of plasma proteins that complement antibodies
- Inflammation
    - Immediate
    - Local
    - Occurs in vascularized tissues
    - Responds against a variety of injury-causing stimuli
- Fever
    - Inhibits replication of bacteria and viruses
    - Promotes interferon activity
    - Accelerates tissue repair

NOTES

Adaptive Immunity

This type of immunity occurs slowly over a period of days. It is a response to a specific pathogenic challenge, usually a bacteria or virus. It is referred to as an immune response. It involves the stimulation of WBCs by antigens on the surface of the invading pathogen. These antigens are protein molecules that the body's immune system recognizes as non-self (not belonging to the human host). The adaptive immune system then produces antibodies, also protein molecules, which attack the foreign invaders by creating antigen-antibody complexes, destroying the pathogen.

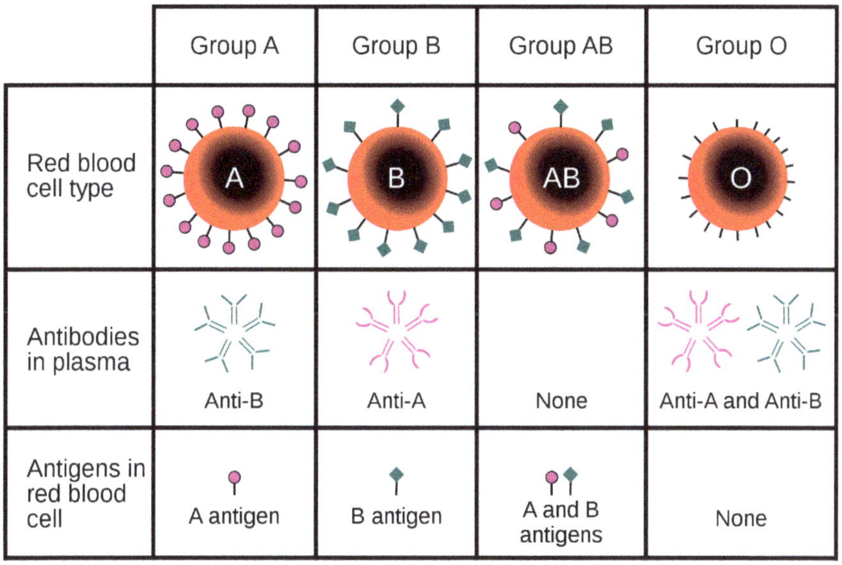

FIG. 15.1 ABO blood typing graphic.

This reaction also applies to the ABO and Rh blood typing systems. RBCs have surface antigens, just like all cells, including bacteria. Our immune systems can recognize whether those antigens are foreign and respond appropriately. The names of blood types are derived from which surface antigens are present. So, type A blood has A surface antigens, B has B antigens, AB has both A and B antigens, and type O has neither A nor B antigens. Someone who is Rh positive has Rh (also known as D) surface antigens, whereas a person who is Rh negative doesn't have Rh antigens. Figure 15.1 shows the presence of surface antigens in the ABO blood types (it does not show the Rh antigen).

If a person has antigens A, for instance, they will not be able to produce the antibody for that antigen. So, if A antigen is present, then no A antibody will be produced. Think about it, if they were type A and produced A antibodies that would destroy their own RBCs!

The eight different blood types are summarized as follows:

- Type AB positive—A, B, and Rh antigens; so, universal acceptor
- Type AB negative—A, B, but no Rh antigens
- Type A positive—A and Rh antigens
- Type A negative—A but no Rh antigens
- Type B positive—B and Rh antigens
- Type B negative—B but no Rh antigens
- Type O positive—only Rh antigens
- Type O negative—no antigens; so, universal donor

✔ **CONCEPT!** The body's adaptive immune system distinguishes between self and non-self, determined by the presence of antigens on the surface of cells. The immune system will respond to non-self antigens by producing antibodies that bind with the non-self (foreign) antigens and destroy the pathogen. Understanding blood typing is a good way to gain insight into adaptive immunity and the interaction of antigens and antibodies.

Figure 15.2 is a flowchart summarizing the concept of adaptive immunity. This figure shows the immune system distinguishing self from non-self, based on surface antigens on the invading pathogen (bacteria or virus). If the immune system detects antigens it recognizes as self, there are no antibodies (ATBs) produced. But if the immune system detects antigens that are foreign, it responds by producing antibodies (ATBs) that will bind to the foreign antigens and destroy the invading pathogen.

✔ **CONCEPT** Rule of two: The adaptive immune system is able to distinguish between self and non-self

There are two types (rule of two) of acquired immunity, active and passive.

- **Active-acquired immunity**—the body's own immune system produces the ATBs.
- **Passive-acquired immunity**—the ATBs are produced by a source outside the body.
- Active immunity involves **endogenous** (made internally) ATBs.
- Passive immunity involves **exogenous** (made outside the body) ATBs.

✔ **CONCEPT!** Active immunity involves the production of endogenous ATBs, whereas passive immunity involves receiving exogenous ATBs.

Discuss the concept of acquired or adaptive immunity in more detail with your instructor and/or classmates. Draw out a chart showing the presence or absence of both surface antigens and antibodies in the eight different blood types. Utilize Figure 15.1, a reference text, or online resources for guidance.

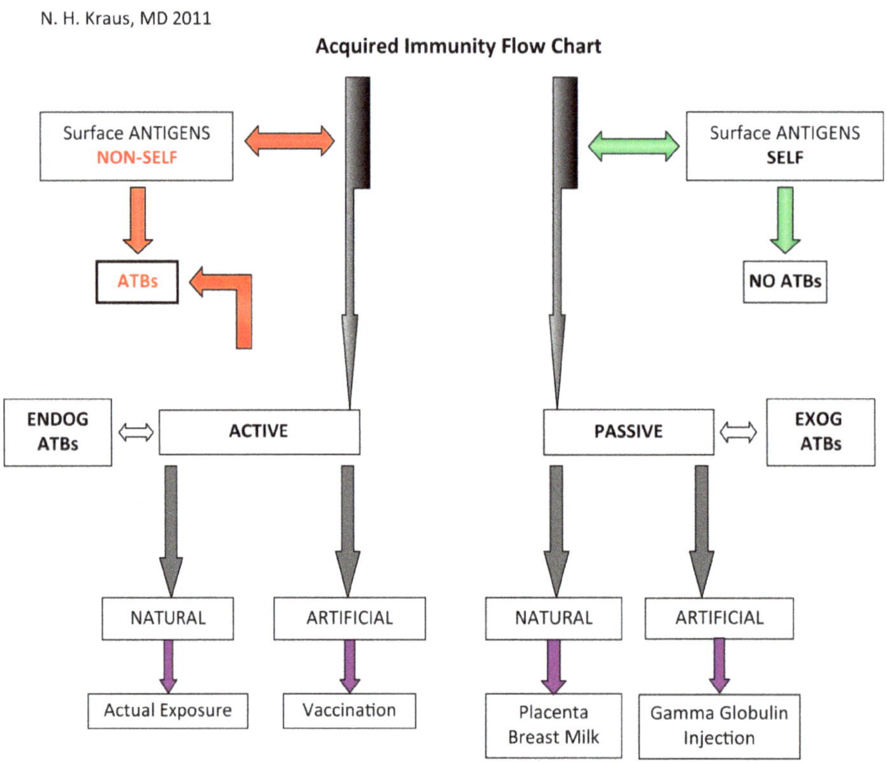

FIG. 15.2 Acquired immunity flowchart.

NOTES

T-Cell and B-Cell Lymphocytes

Immunity is complex and detailed. It is greatly simplified in *SS A&P*. However, students do need to be aware of the two different types of white blood cells which are involved in adaptive immunity. There are two different classes of WBCs, granulocytes and agranulocytes.

When viewed on a histological slide, granulocytes have what appear to be granules in the cytoplasm. Agranulocytes have cytoplasm which appears relatively clear. Lymphocytes are one of the types of agranulocytes and are the principle cell involved in adaptive immunity.

As you might guess (rule of two), there are two types of lymphocytes, T-cells and B-cells. T-cells provide **cellular immunity**. This means the T-lymphocyte must come in direct contact with the pathogen to destroy it, whereas the B-lymphocytes are stimulated by foreign antigens to produce protein antibodies, which will circulate throughout the body, destroying the foreign pathogens. This is called **humoral immunity**.

Summary of cells providing adaptive immunity:

- T-lymphocytes (**T-cells**)—provide cellular immunity; require direct contact with foreign invader
- B-lymphocytes (**B-cells**)—provide humoral immunity; produce circulating antibodies

Primary and Secondary Response in Humoral Immunity

An initial exposure (active infection or vaccination) to a specific antigen creates a **primary response** with the production of IgG and IgM antibodies. This reaction of the immune system will take one to two weeks (lag time). The level of antibodies will decrease over time.

Another exposure to the same antigen will produce a **secondary response**. The levels of antibodies rise more rapidly and to a much higher concentration than with the primary response. Figure 15.3 shows primary and secondary response in graphic format.

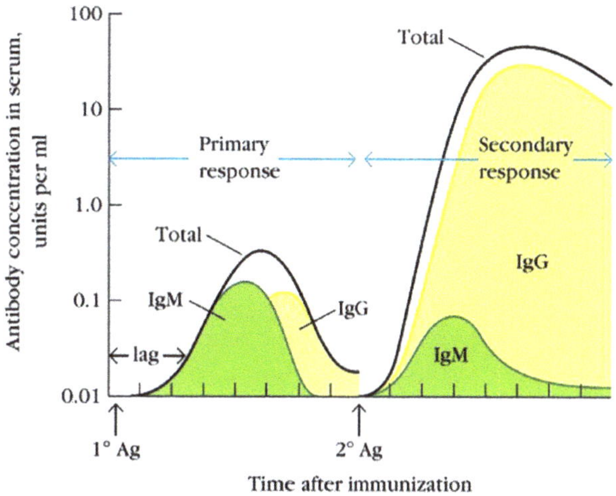

FIG. 15.3 Primary vs Secondary Humoral Response.

Review Questions

- The immune system provides what function for the body?
- What is innate immunity?
- List at least four examples of innate immunity.
- What is acquired immunity, and how does it differ from adaptive immunity?
- What's the difference between endogenous and exogenous antibodies?
- What's the difference between innate and acquired immunity?
- Antigens and antibodies are what kind of molecule?
- Breastfeeding provides what type of immunity?
- A vaccination provides what type of immunity?

Activities and Coloring

- Make a list of all the specific types of innate immunity listed at the beginning of this chapter.
- Explain the acquired immunity flowchart to another A&P student, roommate, friend, or family member.
- Draw out and color a chart showing the antigens and antibodies present in ABO Rh blood typing. Create your own version of Figure 15.1 and add the Rh antigen.

FIGURE CREDITS

Figure 15.1: Source: https://commons.wikimedia.org/wiki/File:ABO_blood_type.svg.

Figure 15.3: Source: https://www.newhealthadvisor.org/images/1HT04389/pis.jpg.

Churn to Learn!

Teach to Learn!

CHAPTER SIXTEEN

The Respiratory System

Gas Exchange

TOES (TOPICS OF EMPHASIS)

- Function of the respiratory system
- The structure of the respiratory system
- Conducting zone—external nares to terminal bronchioles
- Respiratory zone—respiratory bronchioles to alveolar sacs
- The epithelial lining of the respiratory system
- Major muscle of breathing
- Secondary muscles of breathing
- Mechanics of ventilation
- Boyle's law
- Respiratory volumes
- Oxygen and carbon dioxide transport

Respiratory System Provides Gas Exchange

The respiratory system consists of a series of tubes that conduct or transport air from outside the body into the lungs. This is called **ventilation**. Once the air is deep in the lungs, gas exchange takes place, with oxygen going from the alveoli into the pulmonary capillary bed. Carbon dioxide (waste product from cellular respiration) moves from inside the pulmonary capillaries into the alveoli so it can be expired (moved out of the body).

To accomplish ventilation and alveolar exchange of gases, the respiratory system has a conducting (it conducts air) zone running from the external nares (outer opening of the nose) to the terminal bronchioles deep in the lungs. The respiratory zone consists of the respiratory bronchioles, alveolar ducts, alveoli, and alveolar sacs.

The functional and structural unit of the respiratory system is the **alveolus**. These thin-walled sacs are arranged in grape-like clusters, increasing surface area and facilitating the movement of gases through simple diffusion. The epithelial lining of the tubes in the lungs evolves as the function of each area changes.

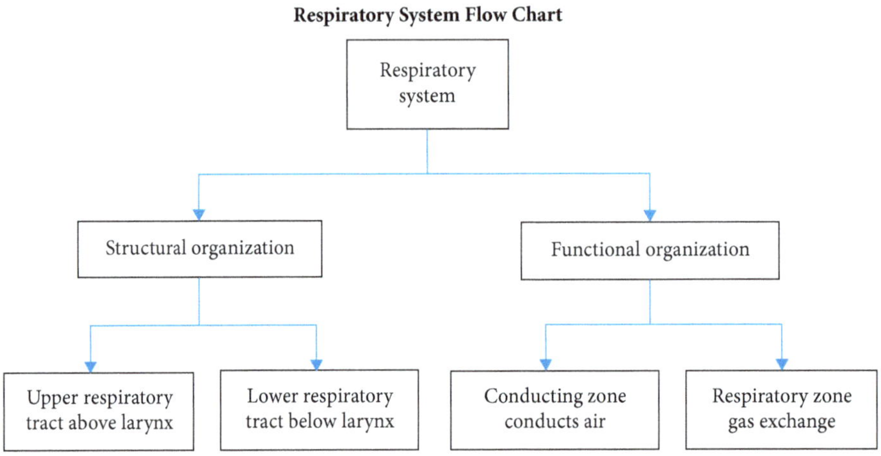

FIG. 16.1 Respiratory system flowchart.

The nasal cavity and larger conducting tubes of the conducting system are lined by pseudostratified ciliated columnar epithelium that protect and clean the lungs. But the tiny tubes of the respiratory zone are lined with simple squamous epithelium to facilitate gas exchange.

Refer to Figure 16.1 showing the flow chart of the respiratory system (RS). Functionally, the RS is organized into two major divisions, the conducting zone (CZ) and the respiratory zone (RZ). The CZ simply conducts air (that's why it's called the conducting zone) from the outside atmosphere into the lungs. The RZ is where gas exchange (oxygen into the CVS and CO_2 out of the CVS) takes place deep within the lungs.

The left and right lungs are similar in structure. But there are differences that the novice A&P student needs to recognize. The left lung has one fissure separating two lobes. It also has a cardiac notch which is a depression where the heart takes up space. The right lung has two fissures and three lobes and no cardiac notch. These allow a student to tell the difference between the left and right lung in lab models.

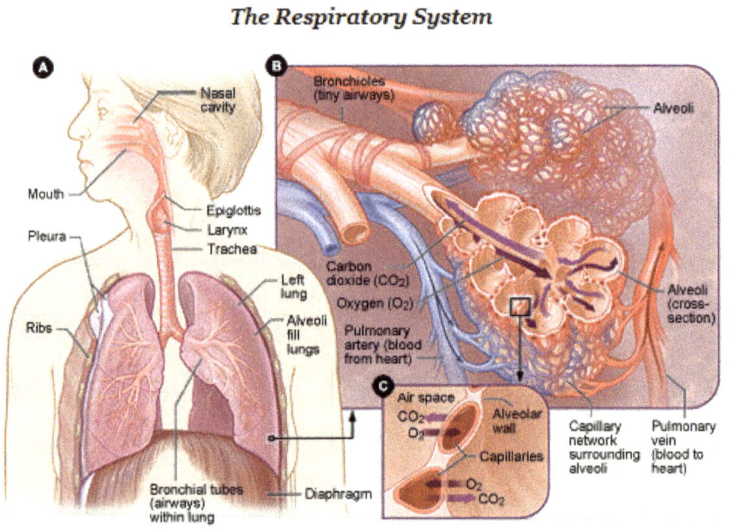

FIG. 16.2 Respiratory system and alveolar structure.

✔ **CONCEPT!** The lungs provide gas exchange, oxygen in and carbon dioxide out. Lungs are designed the way they are because of what they do.

Learn to **run** (**list**) **and identify** the RS structures from the nares to the alveoli, including the following structures:

Conducting zone

- Nares—nostrils
- Nasal cavity
- Nasal pharynx
- Oral pharynx
- Laryngeal pharynx
- Larynx
- Trachea
- Carina
- Primary or main stem bronchi
- Secondary bronchi
- Tertiary bronchi
- Terminal bronchioles

Respiratory zone

- Respiratory bronchioles
- Alveolar ducts
- Alveolar sacs

Muscles of Respiration

The main muscle of respiration is the **diaphragm**. This is a dome-shaped structure appearing a little bit like an upside-down soup bowl. The middle of the diaphragm is dense, regular connective tissue called the **central tendon**. It is a unique tendon since it doesn't connect to any boney structure. It is easy to deduce that the muscle of the diaphragm is made of skeletal muscle because it is voluntary. We can consciously control our breathing.

Figure 16.3 shows the position of the diaphragm when relaxed (blue line) compared to when it contracts (red line). When the diaphragm contracts and flattens it pulls the lungs down and increases the volume inside the lungs. This allows ventilation to occur.

The diaphragm is innervated by the phrenic nerve, with motor input from the cerebral cortex (voluntary) and from respiratory control centers in the brain stem (pons and medulla oblongata). Innervation from the brain stem allows (involuntary) breathing to continue when we sleep or are unconscious, while the motor somatic input allows conscious control of breathing.

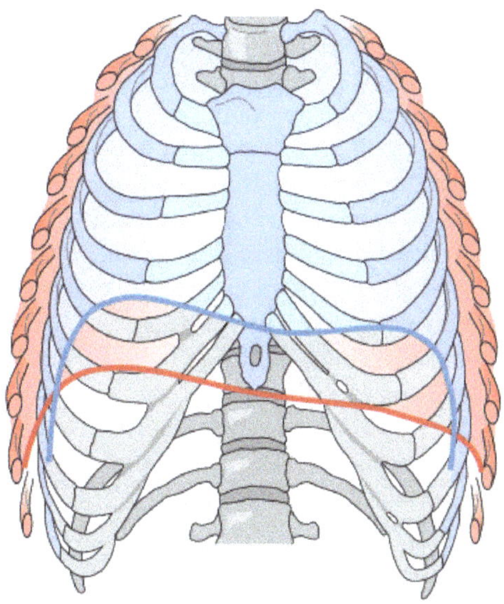

FIG. 16.3 Thoracic cage and diaphragm.

There are secondary muscles of breathing. These include the following:

- Internal intercostals (between the ribs)—involved in forced expiration
- External intercostal—involved in forced inspiration
- Scalene muscles in the neck—involved in forced inspiration
- Abdominal muscles—involved in maximum forced expiration

NOTES

Ventilation

Boyle's law expresses the inverse relationship of pressure and volume and can be stated as a simple formula:

$$P = \frac{1}{V} \ or \ V = \frac{1}{P}$$

✔ **CONCEPT!** When the volume of a closed, gas-filled space increases, the pressure decreases. When the volume decreases, the pressure increases.

Boyle's law is physics but important in human physiology. The formula shows that in a gas-filled container the pressure and volume are inversely related. When one goes up the other goes down. In the respiratory system, the contraction and relaxation of the diaphragm raises and lowers the volume in the pleural cavities, decreasing and increasing the air pressure in the lungs. When the pressure in the lungs drops below the pressure in the surrounding atmosphere, air is literally forced into the lungs by the higher atmospheric pressure.

When we inhale it feels as if we're sucking air into our lungs. But when we understand Boyle's law, and the process of respiratory ventilation, we understand that actually atmospheric pressure is *forcing* air into our lungs. That's why it's more difficult to breathe at higher elevations, like Boulder, Colorado, or on top of a mountain. The atmospheric pressure is lower at higher altitudes, so we work harder to ventilate.

Once again, students are encouraged to go online or use a reference text to gain further insight into Boyle's law and ventilation. Some online sites have animations demonstrating Boyle's law as well as ventilation. These can be very helpful.

Overview of Respiration

Respiration involves four processes that allow the body to get oxygen to the tissues, where it can be used to make energy stored in the form of ATP:

- **Pulmonary ventilation**
  - Oxygen rich air is inhaled
  - Oxygen poor air is exhaled
- **Alveolar gas exchange**
  - Oxygen moves out of alveoli into the blood
  - Carbon dioxide moves out of the blood into alveoli
- **Gas transport through the CVS**
- **Systemic gas exchange**
  - Oxygen moves into cells
  - Carbon dioxide moves into blood

Respiratory Volumes and Pressures

The volume of air that enters and leaves the lungs can be measured with an instrument called a spirometer. Respiratory volumes vary depending on environmental conditions, age, and underlining disease. These volumes can be utilized to help diagnose disease. Figure 16.4 shows these volumes.

The most important respiratory volumes for the novice A&P student to know are tidal volume, vital capacity, and FEV_1:

- **Tidal volume** (TV or V_T)—the volume of air moving in and out of the lungs while at rest. Averages about 500 cc or half a liter of air per breath
- **Vital capacity** (VC)—total amount of air that can be exchanged through one maximum inhalation and exhalation
- **FEV_1**—the amount of air that can be forced out of the lungs in one second. This is usually about 80 percent of the vital capacity. It will decrease in patients with COPD (emphysema and chronic bronchitis).

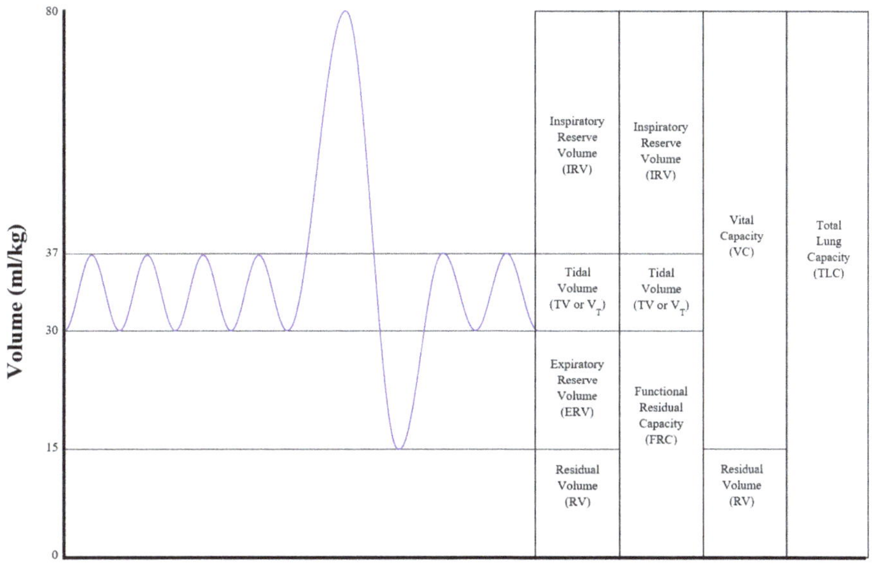

FIG. 16.4 Respiratory volumes and pressures.

NOTES

Gas Exchange

There are two types of gas exchange, the process of gases (oxygen and carbon dioxide), moving from one location to another, and they take place in two areas:

- **Alveolar gas exchange**—takes place in the lungs
- **Systemic gas exchange**—takes place in the tissues (all tissues including lung parenchyma)

Gas exchange is dependent on the partial pressure of gases. Partial pressure is the amount of force or pressure exerted by each gas within a mixture of gases. It is expressed as millimeters (mm) of mercury (Hg) and written with a "P" before the atomic symbol of the gas. For example, the partial pressure of carbon dioxide is written as PCO_2, and PO_2 indicates the partial pressure of oxygen.

A partial pressure gradient exists when the partial pressure of a specific gas (O_2 or CO_2 in this case) is greater in one area than another area. The gas will diffuse from the area of higher concentration to lower concentration. This is simple diffusion.

BRIDGE to Chapter 7, page 70 to review simple diffusion

In the alveolar sacs of the lungs, the partial pressure of oxygen is higher inside the alveolar sacs than the pulmonary capillaries, so oxygen will move down its concentration gradient into the blood vessels. The PCO_2 is greater in the capillaries, so carbon dioxide will move out of the capillaries into the alveolar sacs. Carbon dioxide is then expired (breathed out) during ventilation.

At a tissue level, the partial pressures of oxygen and carbon dioxide are reversed, with PCO_2 higher in the tissues than the capillaries. As a result, carbon dioxide moves into the capillaries. PO_2 in the capillaries is higher than it is in the tissues, so oxygen moves out of the capillaries and into the tissues (cells). Then the cells utilize the oxygen to make energy (ATP) through cellular respiration.

Oxygen and Carbon Dioxide Transport

Oxygen is transported from the lungs to the tissues by the RBCs. In the lungs, hemoglobin (Hb) in the RBCs combines with oxygen to form oxyhemoglobin (HbO_2). At the tissue level, HbO_2 releases the oxygen (dissociates). The oxygen then diffuses down its partial pressure gradient into the cells of the tissue.

Carbon dioxide has three ways of being transported from the tissues to the lungs for expiration:

- CO_2 dissolved in the plasma (like CO_2 in a soda)
- CO_2 bonded to hemoglobin
- HCO_3^- dissolved in the plasma

About 70 percent of the carbon dioxide produced by CR is transported through the RBCs. CO_2 diffuses into the RBCs at the tissue level and combines with water (H_2O) to form carbonic acid (H_2CO_3), which dissociates (breaks down) into bicarbonate (HCO_3^-) and hydrogen ion (H+). The hydrogen ion binds to Hb in the RBCs, and the bicarbonate diffuses out of the RBCs into the plasma. Most of the carbon dioxide in the blood stream is transported back to the lungs as bicarbonate dissolved in the plasma. This reaction can be represented in the following chemical formula:

$$CO_2 + H_2O \leftrightarrow H_2CO_3 \leftrightarrow HCO_3^- + H^+$$

In the lungs, carbon dioxide is regenerated as blood moves through the pulmonary capillaries and the formula is reversed. Notice the arrow indicates the reaction can flow both ways. This is true for most, if not all, chemical reactions in the body: They can flow either way depending on the concentration of reagents and other factors (e.g., presence of enzymes, temperature, etc.).

We'll use this formula again when we discuss acid-base balance and buffering in chapter 20.

Review Questions

- What is the fundamental function of the RS?
- What's the difference between the conducting zone and the respiratory zone?
- What's the difference between internal and external respiration?
- Why are alveoli arranged like a "bunch of grapes"?
- What is the functional unit of the respiratory system?
- What is the structural unit of the respiratory system?
- What is the major muscle of breathing?
- What is the equation for Boyle's law?
- Why is Boyle's law important to understanding ventilation?
- What is tidal volume?
- What does partial pressure mean?

Activities and Coloring

- List the four steps of the respiratory process and explain each process to another student.
- Draw and color a picture of the movement of oxygen and carbon dioxide caused by partial pressure gradients at both the alveolar and systemic levels.
- Explain the respiratory system, both structure and function, to another person who doesn't know A&P. Struggle to teach others to deepen your understanding and insight.
- Run the RS, writing it out on a blank sheet of paper. Do this **repetitively** while saying the structures **out loud** as you write them down. Get so you can rattle this off without hesitation.

NOTES

FIGURE CREDITS

Figure 16.2: Source: https://commons.wikimedia.org/wiki/File:Human_respiratory_system-NIH.PNG.
Figure 16.3: Copyright © by Thieme - General Anatomy and Musculoskeletal System (CC BY-SA 4.0) at https://commons.wikimedia.org/wiki/File:Diaphragm-thieme.png.
Figure 16.4: Copyright © Vihsadas (CC BY-SA 3.0) at https://commons.wikimedia.org/wiki/File:Lungvolumes.svg.

REVV CCCAMM!

Vocalization Is Essential!

Churn to Learn!

Teach to Learn!

CHAPTER SEVENTEEN

The Digestive System

Nutrition and Nutrient Absorption

TOES (TOPICS OF EMPHASIS)

- Divisions of the digestive system
- Structure of the GI tract
- Accessory organs of the GI tract
- Epithelial lining of GI tract
- Muscle layers throughout the alimentary canal
- Conceptual function of the digestive system (DS)
- Nutrients and their digestion

Digestive System

All cells in the human body need energy to function. To create energy, cells need oxygen and nutrients. The three basic nutrients are listed in the H_2O sheet. They are C, P, L, or carbohydrates, proteins, and lipids. Our food contains those three nutrients. The digestive system is designed to ingest food, break the food into smaller pieces, absorb those smaller bits, and circulate them through the CVS to all the cells, where they can be utilized to make energy.

BRIDGE to Chapter 14, Lymphatic System, Lacteals, page 199–200

The digestive system has two main divisions: the alimentary canal or gastrointestinal (GI) tract and the accessory organs. The GI tract is approximately 10 meters in length and runs from the mouth and oral cavity to the anal canal and anus. It is designed to effectively break down ingested food both mechanically (chewing and churning) and chemically (digestive enzymes). The GI tract then facilitates the absorption of the digested food into the blood stream or lymphatic lacteals (large lipid molecules). Do you remember lacteals? If not, take a few minutes and review them at the end of chapter 15. Finally, the colon (large intestine) absorbs water, concentrating waste and storing it until elimination.

The proximal GI tract (oral cavity, pharynx, and esophagus) is lined by nonkeratinized stratified squamous epithelium that provides protection. Then the epithelial lining changes very suddenly at the gastroesophageal junction to simple column epithelium with microvilli. The microvilli help to increase the surface area of the lining of the GI tract necessary to facilitate absorption of nutrients. The epithelium lining the mucosa of the GI tract changes because its function changes. **FFF!**

✔ **CONCEPT** Rule of two: Notice the DS (digestive system) has two major divisions. We see this pattern repetitively. A system or structure has two divisions. Each of those two divisions has two subdivisions. Use the rule of two to your advantage as you CTL (churn To learn) A&P.

Just as with the respiratory system, the novice A&P student needs to be able to run (list) and identify the structures and organs of the GI tract from the mouth to the anus including the following structures:

- Mouth—this is an opening into the oral cavity; epithelium lining proximal parts of the GI tract is nonkeratinized stratified squamous
- Oral cavity—contains the teeth and tongue
- Oral pharynx
- Laryngeal pharynx
- Esophagus
- Gastroesophageal junction—contains the cardiac sphincter (located just inferior to the heart); epithelium changes to simple columnar with micovilli
- Stomach—fundus, body, and pylorus areas
- Pyloric sphincter
- Duodenum
- Jejunum
- Ileum
- Ileocecal sphincter
- Cecum—appendix
- Ascending colon
- Hepatic flexure
- Transverse colon
- Splenic flexure
- Descending colon
- Sigmoid colon
- Rectum
- Anal canal—epithelium converts back to nonkeratinized stratified squamous
- Anus—an opening (internal and external anal sphincters)

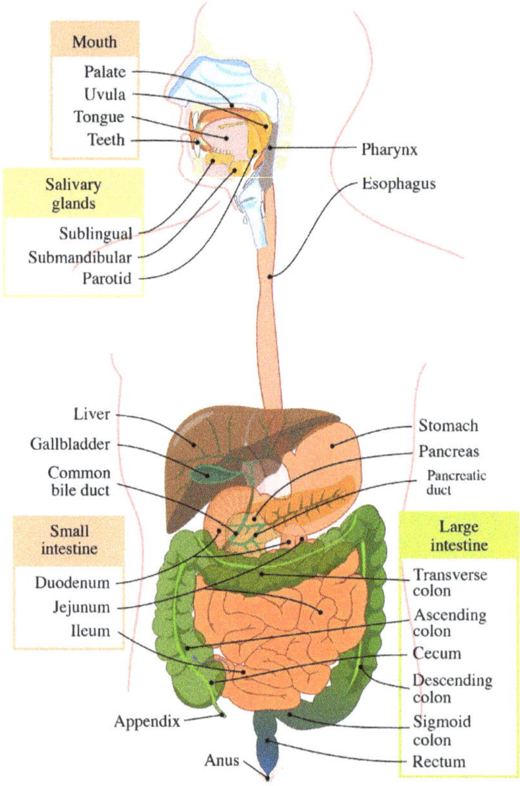

FIG. 17.1 GI tract from mouth to anus.

Sphincters

Sphincters are found throughout the body, including the digestive, urinary, and cardiovascular systems. Sphincters are circles of muscle, usually visceral but skeletal in two areas, which surround tubular structures (e.g., arterioles, anal canal, and urethra). They contract and relax, controlling the flow of material (blood, feces, urine) through the tube.

The GI tract has numerous sphincters (e.g., cardiac sphincter, pyloric sphincter, sphincter of Oddi, etc.). The internal and external anal sphincters control the storage and release of feces from the rectum through the anal canal. There are similar internal and external sphincters in the urinary system.

As the rectum fills with feces, sensory receptors in the wall of the rectum begin a spinal cord reflex arc, which causes the involuntary relaxation of the internal anal sphincter (visceral muscle). This creates an urge to defecate.

But, fortunately, the external anal sphincter is skeletal muscle under voluntary control. Therefore, an individual can hold the feces until in an appropriate place to defecate.

In addition to listing and identifying structures in the GI tract, the introductory student also needs to learn the six accessory organs of the digestive system. Those accessory organs are not directly a part of the alimentary canal but augment or support the function of the GI tract mechanically or chemically. Those accessory organs are as follows:

BRIDGE
to Chapter 12, Reflex Arcs, page 165

BRIDGE
to Chapter 18, Urinary System, page 239 for discussion of urethral sphincters

- Tongue
- Teeth
- Salivary glands
- Liver
- Gall bladder
- Pancreas

The pancreas produces all three digestive enzymes. Amylases breakdown carbohydrates, proteases catabolize proteins, and lipases breakdown fats. The chemical breakdown of carbohydrates starts in the oral cavity, digestion of proteins in the stomach, and fats in the small intestine. Remember, in the H_2O sheet the primary nutritional biochemicals are listed C, P, L, reflecting their order of digestion.

H_2O

The two muscle layers (one longitudinal, the other circular) making up the muscularis layer of the GI tract churn and mix the food as it is digested. The stomach has a third oblique layer, allowing that organ to fulfill a major function of mixing and churning food once it passes out of the esophagus into the stomach. *This is one more example of FFF!*

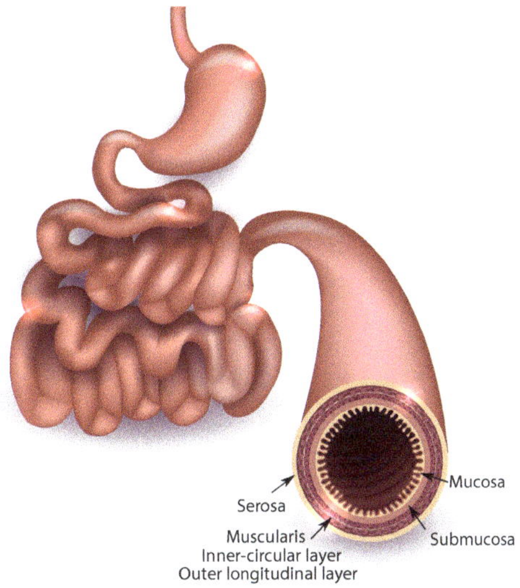

FIG. 17.2 Cross section of small intestine showing layers of wall.

The small intestine has three specific characteristics that help increase mucosal layer surface area, facilitating absorption of nutrients. Figures 17.2 and 17.3 show these features of the small intestine.

- Villi
- Microvilli covering the surface of simple columnar cells, which cover the villi
- Circular folds

The large intestine also has three distinguishing characteristics. Figures 17.4 and 17.5 demonstrate these characteristics.

- Haustra
- Teniae coli
- Omental or epiploic appendices

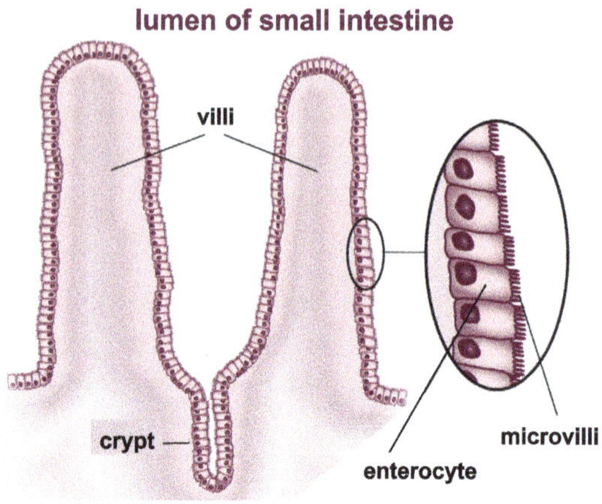

FIG. 17.3 Villi and microvilli lining the lumen of small intestine.

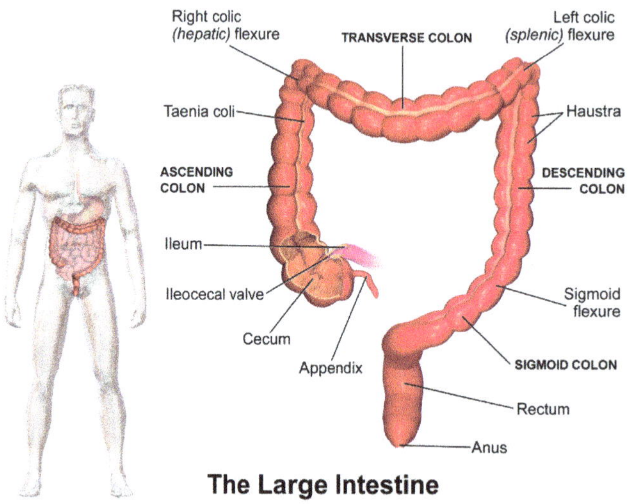

FIG. 17.4 Areas and characteristics of large intestine.

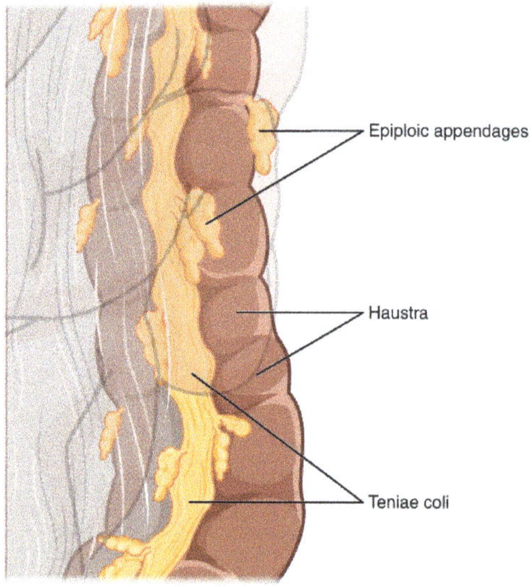

FIG. 17.5 Characteristics of large intestine.

NOTES

Divisions of the DS

Let's take a more detailed look at the two divisions of the digestive system as well as the breakdown of nutrients.

- Alimentary canal or the gastrointestinal (GI) tract
  - Tube that is approximately 10 meters long
  - Mouth and oral cavity proximally to anal canal and anus distally
  - Nutrients are ingested, broken down, and absorbed in G-I tract
    - CHO
    - Proteins
    - Lipids
  - Colon concentrates and eliminates
- Accessory organs
  - Accessory organs augment the GI tract functions
  - Nutritents are broken down more completely by accessory organs
  - ***Accessory organs in an acronym* - T, T, S, L, G, P**
    - **T**eeth
    - **T**ongue
    - **S**alivary glands
    - **L**iver
    - **G**all bladder
    - **P**ancreas

Nutrient Breakdown

- **Carbohydrates** (CHO)—polysaccharides break down into monosaccharides
  - Glucose
  - Major source of energy production (CR)
- **Proteins**—polypeptides break down into amino acids
  - Building blocks for protein synthesis
  - Tertiary source for energy production
- **Lipids**—break down into fatty acids
  - Building blocks for lipid-based molecules (e.g., steroid-based hormones)
  - Secondary source for energy production

Phases of Swallowing

Nutrients taken into the mouth through the oral cavity are broken down both mechanically (chewing) and by saliva (amylase). This creates a bolus of food, which is then swallowed. There are three phases of swallowing.

- **Voluntary** phase—bolus (ball of partially digested food) is pushed by tongue against hard palate toward the oropharynx
- **Pharyngeal** phase—bolus moves into oropharynx
  - Epiglottis covers the laryngeal opening preventing food from entering the trachea
  - Bolus moves into the esophagus
- **Esophageal** phase
  - Peristaltic contractions of visceral muscle
  - Bolus pushed toward and then into stomach

Gastric Secretions

The lining (mucosa) of the stomach is filled with gastric pits. There are three primary cells in the gastric pits of the stomach, mucous cells, parietal cells, and chief cells.

- **Mucous neck cells**—produce **mucus** (that's why they're called mucous neck cells!)
  - Help to protect the stomach lining (mucosa) against the acidity of HCl
  - Provide lubrication helping the stomach churn and mix the digesting food
  - Produce mucus—different spelling of mucous (glands) and mucus (secretion)
- **Parietal cells**—produce two substances (rule of two)
  - **Intrinsic factor**—essential product!
    - Factor essential for the absorption of vitamin B12
    - Vitamin B12 necessary for production of normal RBCs
    - Intrinsic factor deficiency causes pernicious anemia
  - **HCl**—hydrochloric acid
    - Acid is produced on the surface of the cell, not inside the cell
    - Production inside the cell would destroy the cell
    - HCl then interacts with pepsinogen from chief cells to produce pepsin (protease)
    - Pepsin denatures (modifies) proteins
    - HCL kills bacteria—part of innate, barrier immune system
- **Chief cells**—most numerous of gastric pit glands
  - Cells produce pepsinogen—precursor of pepsin
  - Pepsinogen interacts with HCl to produce pepsin

NOTES

Regulation of Digestion in the Stomach

There are three phases of gastrointestinal digestion:

- **Cephalic**
  - Initiated by thought, smell, sight of food
  - Gets the "juices flowing"
- **Gastric**
  - Initiated when food enters the stomach
  - Causes release of gastrin (paracrine or local hormone)—stimulates stomach glands to secrete digestive enzymes
- **Intestinal**
  - Initiated by presence of acidic chyme (partially digested food) in the small intestine
  - Release of CCK (cholecystokinin)—stimulates bile secretion

Digestion and Absorption

The function of the DS is the breakdown, both mechanical and chemical, and then the absorption of essential nutrients. Although digestion of nutrients can be complex, we'll simplify the process to build a foundation of understanding. The body's essential biochemicals, C, P, and L, found in column two of the H_2O sheet, are listed in the order of their chemical digestion, starting in the oral cavity. Nucleic acids contained in both plant and animal-based foods do not make up a large component of food nutrients and so are ignored at a novice level of learning A&P.

- **Carbohydrates** (**CHO**)
  - Start chemical digestion in the oral cavity
  - Salivary amylase secreted by salivary glands
  - Continue breakdown in small intestine
  - Pancreatic and intestinal amylases finalize digestion

- **Proteins**
  - Begins in the stomach
  - HCl denatures (modifies) proteins
  - Pepsin further breaks down proteins
  - Pancreatic and intestinal proteases contribute the digestion
- **Lipids (fats)**
  - Bile is secreted by the gall bladder in response to CCK production in the stomach
  - Bile emulsifies fat globs—breaking large fat globules down into smaller fat droplets, facilitating the action of pancreatic and intestinal lipases by increasing the surface area of fat available for enzyme action
  - Pancreatic and intestinal lipases finalize digestion of lipids

Rhythmical contraction of the circular and longitudinal muscles (visceral muscle) in the intestine creates **peristalsis,** which moves the bolus of chyme along the GI tract. Absorption of nutrients continues throughout the small intestine, which is lined by simple columnar epithelium with microvilli. Microscopically, this layer of microvilli looks like the bristles in a brush and so is sometimes referred to as the "**brush border.**" The microvilli on the columnar cells covering the villi in the mucosa of the small intestine increase surface area, facilitating absorption of nutrients.

BRIDGE to Chapter 7, page 68 to review cell surface modifications

Peristalsis continues to move the chyme along through the small intestine and into the colon, passing through the ileocecal sphincter. The main functions of the colon are the resorption of water and the concentration of waste. The waste is held in the rectum. As waste accumulates the rectum stretches. The stretching triggers sensory receptors in the wall of the colon, causing a spinal cord reflex and relaxation of the internal anal sphincter (visceral or involuntary muscle). This involuntary relaxation creates the urge to defecate. However, the external anal sphincter is made of skeletal muscle and so it is under voluntary control. This allows the individual to control defecation until the appropriate time and place.

There is a similar arrangement of internal and external sphincters in the urinary system.

Review Questions

- What is the function of the digestive system?
- Why is the GI tract shaped as it is?
- What are the divisions of the digestive system?
- Why does the stomach have three muscle layers?
- What type of epithelium lines the proximal portion of the GI tract?
- What epithelium lines most of the GI tract?
- What are the six accessory organs of the digestive system?
- What nutrient provides the building blocks for protein synthesis?
- What is gastrin?
- What is CCK?
- Are nucleic acids considered one of the essential nutrients?
- What are the three distinguishing characteristics of the small intestine?
- What are the three distinguishing characteristics of the large intestine?
- What is the order of nutrient digestion?

Activities

- Discuss the conceptual function of the DS with another student.
- Write out a schematic diagram of the GI tract. Color the different organs and areas of the alimentary canal and accessory organs. Explain your diagram to another student, describing the function of all the organs of the DS, including the accessory organs.
- On a blank sheet of paper write out the entire sequence of the GI tract. In other words, run the GI tract from mouth and oral cavity to the anal canal and anus.
- List the six accessory organs and their contribution to the digestive system

NOTES

FIGURE CREDITS

Figure 17.1: Source: https://commons.wikimedia.org/wiki/File:Digestive_system_diagram_en.svg.
Figure 17.2: Copyright © 2016 Depositphotos/Megija.
Figure 17.3: Copyright © BallenaBlanca (CC BY-SA 4.0) at https://commons.wikimedia.org/wiki/File:Villi_%26_microvilli_of_small_intestine.svg.
Figure 17.4: Copyright © by BruceBlaus (CC BY 3.0) at https://commons.wikimedia.org/wiki/File:Blausen_0604_LargeIntestine2.png.
Figure 17.5: Copyright © OpenStax College (CC BY 3.0) at https://commons.wikimedia.org/wiki/File:2433_Teniae_Coli_Haustra_Epiploic_Appendage.jpg.

Vocalization Is Essential!

Churn to Learn (CTL)!

Teach to Learn!

CHAPTER EIGHTEEN

The Urinary System

Waste and Balance

TOES (TOPICS OF EMPHASIS)

- The function of the urinary system
- Organs and structure of the urinary system
- Relating the function of the urinary system to its structure
- Three basic physiological processes of urine production

Urinary System

The urinary system cleans (filters) waste from the blood stream and balances a wide variety of stuff (e.g., electrolytes, water, acidity, glucose, etc.) in blood. It is structured to facilitate those two functions. This is another example of FFF. Organs are structured the way they are because of what they do.

There are two kidneys located in the muscles of the back just behind the parietal peritoneum of the abdominal cavity (retroperitoneal). Figure 18.2 shows the retroperitoneal location of the kidney buried deep in the back muscles, posterior to the abdominal cavity. The introductory student does not need to be concerned with the detailed labels in Figure 18.2. Although the names of some structures will tell you about that structure. For instance, can you figure out what the "perinephric space" is? Learning how to figure things out is an essential part of learning to understand A&P. What is

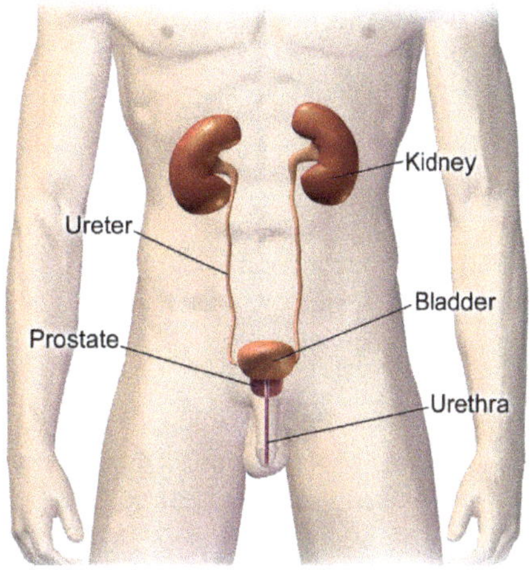

FIG. 18.1 Organs of the male urinary system.

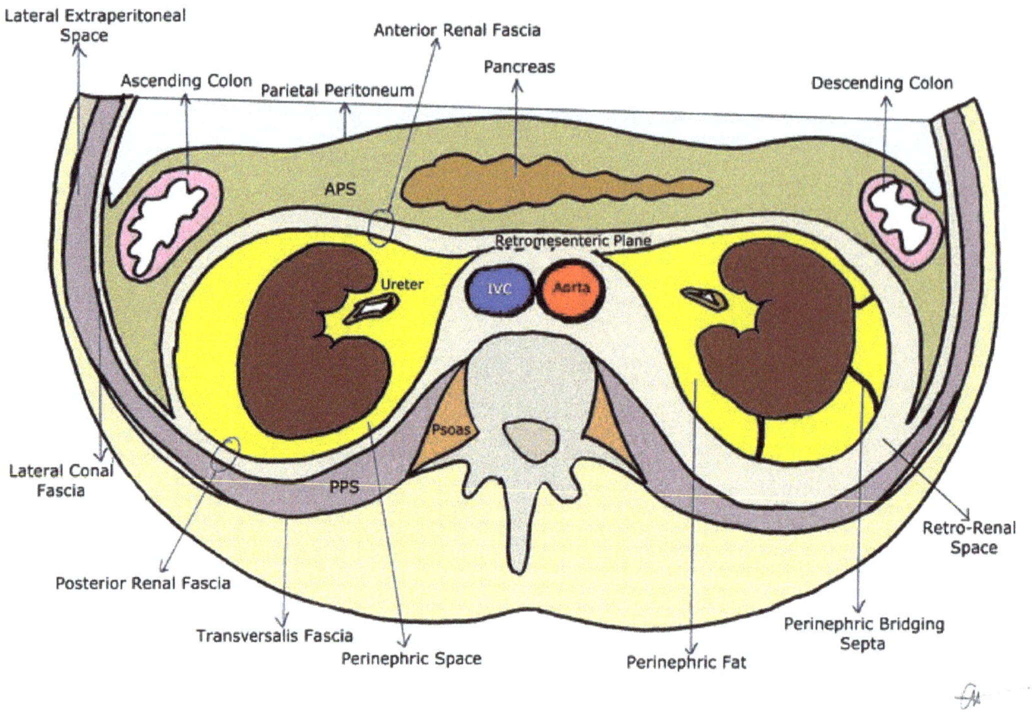

FIG. 18.2 Retroperitoneal location of kidneys.

important is for the student to grasp the concept of the retroperitoneal location of the kidney.

An introductory A&P student needs to be able to identify the following gross structures of the renal system:

- Kidney(s) (2)
- Ureter(s) (2)
- Urinary bladder
- Urethra

Figure 18.3 shows a frontal or coronal section demonstrating the major structures inside the kidney.

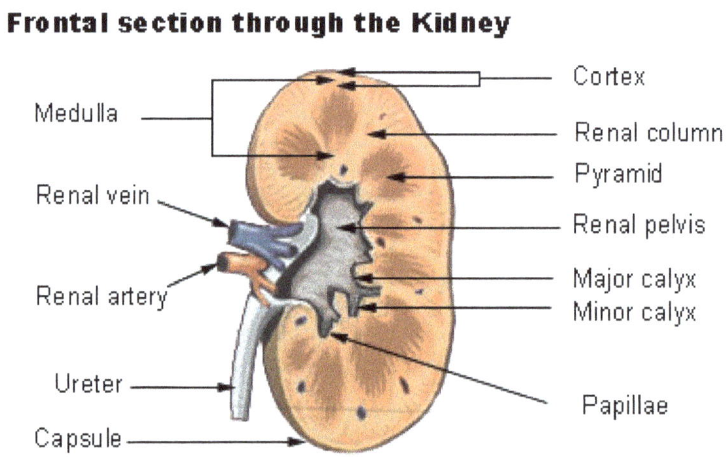

FIG. 18.3 Frontal section showing major structures of the kidney.

The wall of the urinary bladder is composed of visceral muscle called the **detrusor muscle**. It is involuntary muscle that contracts through a spinal cord reflex in response to the bladder stretching as it fills with urine and in response to relaxation of the external urethral sphincter. The mucosa of the bladder is lined by transitional epithelium, which has the unique ability to stretch as the bladder fills.

Students need to learn how to run (list) the structures of the urinary system that are involved in urine production. The need-to-know structures in this sequence are as follows:

- Renal corpuscle—glomerulus + Bowman's capsule (glomerular capsule)
- Proximal convoluted tubule (PCT)
- Loop of Henle—descending and ascending segments
- Distal convoluted tubule (DCT)
- Collecting duct
- Renal pyramid
- Renal papillae
- Minor calyx
- Major calyx
- Renal pelvis
- Ureter(s)
- Urinary bladder—transitional epithelium
- Urethra(s)—internal and external urethral spincters

The **nephron** is the structural and functional unit of the urinary system. It is made up of the following structures:

- Renal corpuscle
- PCT
- Loop of Henle
- DCT

The nephron drains the urine into the collecting duct that drains through the renal papillae into the minor calyx, major calyx, renal pelvis, ureters, urinary bladder, and finally the urethra with release to the outside.

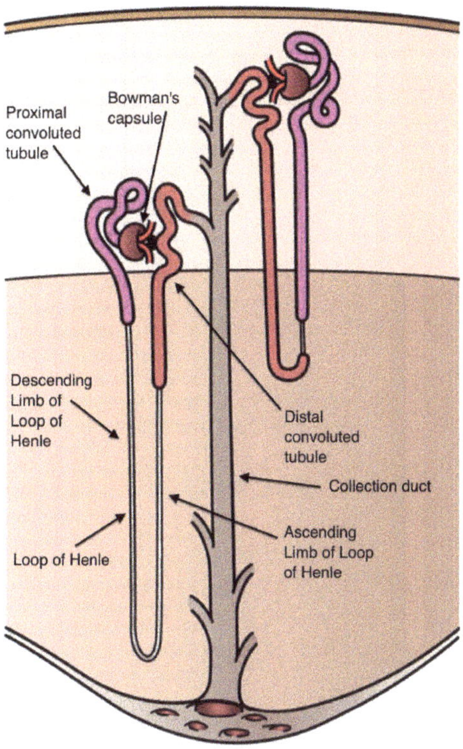

FIG. 18.4 Nephron, the functional unit of the kidney.

NOTES

The urinary or renal system is an interface between the CVS and the renal ducts of the kidney. The kidneys are a miraculous organ that are structured the way they are because of what they do. They filter waste from the blood and balance a bunch of substances. Those substances include water, hydrogen ions, sodium, chloride, potassium and calcium ions, glucose, and more.

It is essential for the novice A&P student to understand the conceptual function of the urinary system. It filters and balances! A closer look at how the renal system filters, balances, and resorbs material is important. But renal function can be extremely complex and overwhelming with osmotic pressure and counter-current multipliers. So, as we've been doing, we'll focus on simplistic concepts, allowing students to practice REVV CCCAMM while building a foundation of understanding.

✔ **CONCEPT!** The renal system filters wastes from the blood and balances the blood levels of a variety of substances!

There are four steps to urine formation and excretion that are essential for the student to grasp:

- **Filtration**
- **Reabsorption**
- **Secretion**
- **Excretion**

The kidneys filter approximately 180 liters of liquid from the blood into the glomerular capsule every 24 hours. That's a little more than 47 gallons! But, humans only urinate approximately 1 to 2 liters a day. Let's take an average and say humans void 1.5 liters each day. So, what happens to all those other liters (178.5)? Where do they go?

The kidneys reabsorb all but 1.5 liters of what it filters, producing a concentrated urine. In addition, small amounts of substances (H^+, calcium, etc.) are secreted into the renal duct system, allowing the kidneys to completely balance the homeostasis of blood and body fluids. The **kidneys** are the primary organs that **regulate fluid balance** in the body.

Filtration

Water and solutes are forced from the glomerulus, or glomerular capillary bed (CVS), into the glomerular (Bowman's) capsule because of the pressure differences across the filtration membrane. The glomerulus, together with the glomerular capsule, is called the **renal corpuscle**. Figure 18.5 shows the renal corpuscle, including the filtration membrane and the podocytes that wrap around glomerular capillaries but do not surround them completely.

The filtered fluid is called filtrate, which is essentially protein-free plasma. So, that 180 liters a day is really plasma, just without the proteins. There are three opposing forces creating the pressure to filter the blood into the glomerulus.

- **Glomerular hydrostatic (blood) pressure (HPG)**—60 mmHg out of capillaries
  - HPG is the driving force that pushes the filtrate out of the glomerular capillaries and into the glomerular capsule

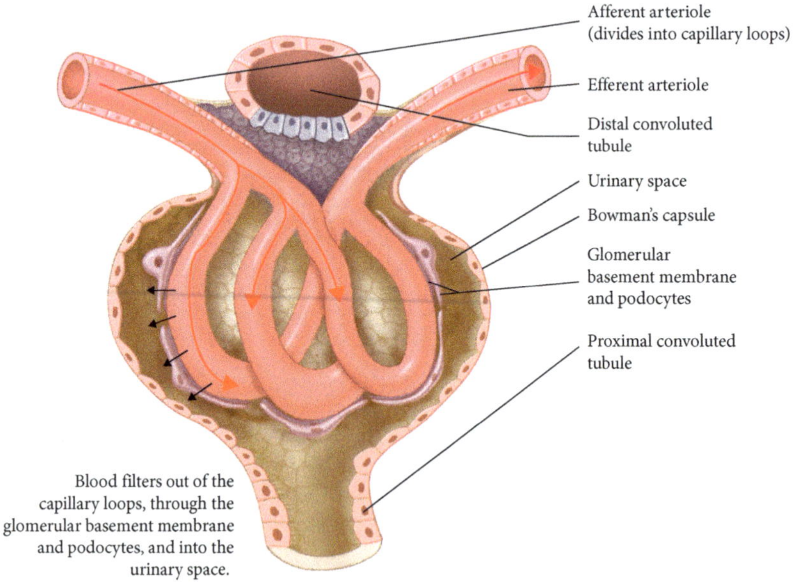

FIG. 18.5 Renal corpuscle.

- HPG is higher than the blood pressure of other systemic arteries
  - The increased pressure is caused by size difference in afferent and efferent arterioles
  - The afferent arteriole is much larger than efferent
  - The afferent arteriole delivers a volume greater than the volume going out the efferent arteriole
  - The volume difference creates pressure which drives filtration
- **Blood osmotic pressure (OPG)**—32 mm Hg into capillaries
  - Osmotic pressure is created by the solutes dissolved in blood. Those ismotically active solutes suck, remember?!
  - Solutes are mostly plasma protein molecules
  - The solutes draw fluid (water) back into the glomerular capillaries
- **Capsular hydrostatic pressure (HPC)**—18 mm Hg out of glomerulus and into capillaries
  - Blocks the flow of additional filtrate into Bowman's capsule
  - Opposes filtration

Doing the math shows 60 – 32 – 18 = +10. So, 10 mmHg pressure is creating filtrate.

The **glomerular filtration rate (GFR)** is a standard measurement of the amount of filtrate created in one minute. It is affected by the three opposing forces (HPG/OPG/HPC) discussed above.

NOTES

Reabsorption

This is where the kidneys do an enormous amount of their work. Most of the filtrate must be reabsorbed as it passes through the renal tubules of the nephron. Remember, the nephron is the structural and functional unit of the renal system. The substances reabsorbed can be categorized into three groups:

- **Substances or stuff that are completely reabsorbed**
  - One hundred percent of this stuff is reabsorbed
  - The PCT (proximal convoluted tubule) resorbs the majority of these substances

- Two major types of stuff in this group (rule of two again!)
  - Nutrients—**glucose**, amino acids, lactate
  - Plasma proteins—only small amount filters and all reabsorbed

- **Substances with regulated reabsorption**
  - Some but not all these substances are taken back into the peritubular capillary bed
    - The reabsorption of sodium (Na^+) is of great importance here. Amount of sodium reabsorbed varies between 95 and 100 percent
    - Reabsorption occurs along the entire length of the nephron tubes
    - **Aldosterone** (adrenal cortical endocrine hormone) is the primary hormone regulating **sodium reabsorption**
    - Many other hormones influence sodium reabsorption
      - Insulin
      - Glucagon
      - Estrogen
      - Progesterone
      - Thyroid hormone
  - Water reabsorption—occurs by osmosis
    - Water reabsorption depends on the state of hydration
    - Reabsorption of water follows the osmotic gradient of sodium concentration (solutes suck!)
    - **ADH** (**antidiuretic hormone**) produced in the hypothalamus, stored and secreted by the posterior pituitary is primary hormone regulating water reabsorption
  - Potassium—both reabsorbed and secreted
  - Calcium and phosphate
  - Hydrogen ions and bicarbonate

- **Substances eliminated as waste products**
  - Nitrogenous waste
    - Urea—produced from protein breakdown in the liver
    - Uric acid—produced from nucleic acid breakdown in the liver
    - Creatinine—produced from the metabolism of creatine in muscle
  - Certain drugs—eliminated by both filtration and secretion
    - Antibiotics—penicillin, sulfonamides
    - Aspirin
    - Chemicals in marijuana
  - Other metabolic wastes

- Some hormones
    - HCG (human chorionic gonadotrophin)
        - Produced by implanted zygote (fertilized egg) in early pregnancy
        - Used to verify pregnancy (urine pregnancy test)
    - Epinephrine

Secretion

Substances such as potassium, hydrogen ions, and bicarbonate are substances the kidneys filter, reabsorb, and secrete. Secretion occurs in the renal tubules and allows the kidneys to balance a variety of materials.

Excretion

Finally, urine is stored in the urinary bladder. As it becomes full, stretch receptors in the wall of the bladder begin a spinal cord reflex arc, which causes the relaxation of the internal urethral sphincter muscle. This is a circle of visceral (involuntary) muscle that surrounds the proximal urethra.

BRIDGE to Chapter 17, Sphincters, page 224

This results in the urge to urinate we all experience. But, just as in the anal canal, there is an external urethral sphincter. This is made of straited muscle and is under voluntary control, so the urine can be held until an appropriate opportunity to void.

The two major processes involved in kidney function are filtration and reabsorption, with reabsorption being the most complex part of kidney function. We've stayed in the basement again and not addressed some of the complexity of renal function like the counter current multiplier! That's first- and second-story stuff you can learn (if needed) as you progress in your studies and career.

Figure 18.6 shows a simplified graphic of the phases of renal function. This is the essence of what the introductory student needs to "take away" (TAs) with them regarding the function of the urinary system.

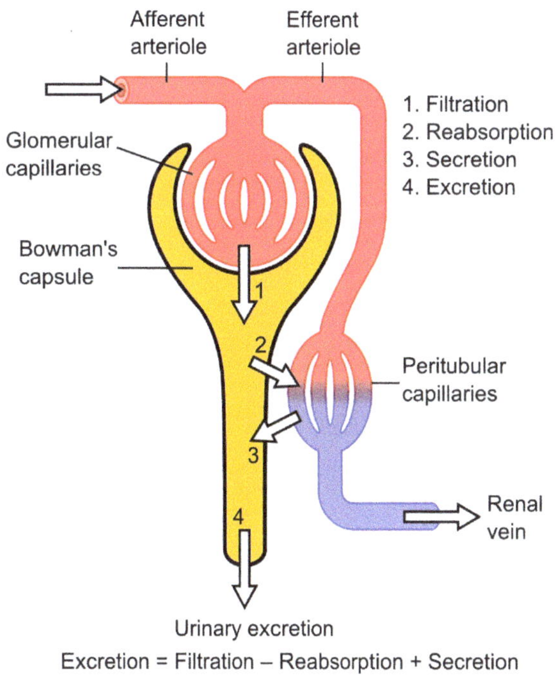

FIG. 18.6 Conceptual graphic of kidney function.

Review Questions

- Conceptually, what does the renal system do?
- What are the major organs of the urinary system?
- What is the functional and structural unit of the urinary system?
- What are the four processes involved in urine production?
- What are the three pressures creating GFR?
- Sodium reabsorption is primarily controlled by what hormone?

- What is HCG?
- Water homeostasis is controlled by what endocrine hormone?

Activities and Coloring

- Draw out and color a diagram of renal function showing the four phases of kidney function, filtration, reabsorption, secretion, and excretion.
- Simplistically discuss the conceptual function of the renal system with another student, friend, roommate, or family member.
- On a blank sheet of paper write out the sequence of urine formation, from the glomerulus to the urethra. Do this *repetitively* while saying the structures *out loud!*
- Draw and color a diagram of the nephron (structural and functional unit of the kidney) in the space provided.

NOTES

FIGURE CREDITS

Figure 18.1: Copyright © by BruceBlaus (CC BY-SA 4.0) at https://commons.wikimedia.org/wiki/File:UrinarySystemMale.jpg.
Figure 18.2: Copyright © by Goran Mitreski and Tom Sutherland (CC BY-SA 4.0) at https://commons.wikimedia.org/wiki/File:Retroperitoneal_spaces_and_fasciae.jpg.
Figure 18.3: Source: https://commons.wikimedia.org/wiki/File:Illu_kidney2.jpg.
Figure 18.4: Copyright © Holly Fischer (CC BY 3.0) at https://commons.wikimedia.org/wiki/File:Kidney_Nephron.png.
Figure 18.5: Copyright © Tieum (CC BY-SA 4.0) at https://commons.wikimedia.org/wiki/File:Glomerular_Physiology.png.
Figure 18.6: Copyright © by Madhero88 (CC BY 3.0) at https://commons.wikimedia.org/wiki/File:Physiology_of_Nephron.png.

Vocalization Is Essential!

Churn to Learn!

Teach to Learn!

CHAPTER NINETEEN

Nutrition and Metabolism

TOES (TOPICS OF EMPHASIS)

- Nutrients
- Regulating nutrient blood levels
- Cholesterol
- Lipid transport
- Conceptual metabolism

Nutrition

Nutrition is the study of how living organisms obtain and utilize nutrients to sustain life. Nutrients are the large biochemical macromolecules listed in column two of our H_2O sheet.

BRIDGE to chapter 6 protein synthesis, page 62

- **Carbohydrates (CHO)**—sugars, starches, fiber
  - Polysaccharides break into disaccharides, and then into monosaccharides
  - Glucose—monosaccharide; major source of energy production
- **Proteins**—most structurally and functionally diverse nutrient molecules
  - As many as one hundred thousand proteins in the body
  - Synthesized from twenty amino acids (two [surprise!] divisions)
  - Essential amino acids (eight) must be obtained from the diet
  - Nonessential amino acids (twelve) can be synthesized by cells
  - Utilized by cells in protein synthesis
- **Lipids (fats)**—diverse group of hydrophobic (insoluble in water; don't mix with water!) molecules
  - **Triglycerides**—most common fat molecule in body; stored in fat tissue
  - **Saturated**
    - Molecule does not contain any double bonds
    - Molecules are completely saturated with hydrogen atoms

    - Saturated fats are solid at room temperature
    - Butter, fat in meat
  - **Unsaturated**
    - Molecule has one double bond between carbons atoms
    - Molecules are not completely saturated with hydrogen
    - Liquid at room temperature (usually)
    - Vegetable oils, canola oil
  - **Polyunsaturated**
    - Two or more double bonds between carbon atoms in the molecule
    - Liquid at room temperature
    - Soybean oil, corn oil, safflower oil
  - **Phospholipids** (remember the SPPLBL?)
  - **Steroids**—including cholesterol
    - Cholesterol is necessary for life
    - Makes up about 20 percent of the SPPLBL (cell membrane)
    - Needed for the synthesis of steroid hormones
      - Three adrenal cortical hormones: aldosterone, cortisol, androgenic hormone
      - Sex hormones: estrogen, progesterone, testosterone
    - **Most of the cholesterol in our blood stream is made in our own liver!**
    - Hepatic synthesis of cholesterol is primarily driven by the dietary intake of saturated fats (butter, marbled meats, etc.)

In addition, nutrients include vitamins and minerals needed to maintain anatomical structures and physiological processes. Of course, water (H_2O) is also a necessary nutrient. The average person requires 2 to 3 liters of water intake each day.

We obtain these nutrients in our food and drink. Nutrients can be described in two ways, each of which has two subcategories. There it is again! There are two different ways to describe nutrients, and each of those has two divisions! **Rule of two, again!**

✔ **CONCEPT** Rule of two: There are two types of nutrients - Macro and Micro

---

Two **types** of nutrients:

- **Macronutrients**—organic biochemicals
  - Body must obtain these in large quantities
  - Carbs, proteins, and lipids
- **Micronutrients**—smaller molecules
  - Vitamins
  - Minerals

Nutrients are also divided into two **categories**:

- **Essential** nutrients—must be obtained through the digestive system
- **Nonessential** nutrients—produced by biochemical processes in the body

Two of the most important nutrients are glucose and calcium. We've already discussed these in detail earlier in *SS A&P*, but they're worth a brief revisit and review. Glucose is a monosaccharide derived from carbohydrates or starches. Calcium is an essential ion found in milk, cheese, spinach, salmon, soybeans, perch, rainbow trout, and other dietary sources.

BRIDGE to chapter 10, page 126 and chapter 13, page 183 to review the homeostasis of blood calcium and glucose.

Lipid Transport

Lipids are hydrophobic (don't like water) and insoluble in plasma. Therefore, lipid transportation requires that the lipids are wrapped in a water-soluble protein. These protein lipid wraps are called **lipoproteins.**

Different types of lipoproteins:

- **Chylomicrons**—necessary for absorption of fats from the small intestine
  - Formed in epithelial cells lining the small intestine
  - Composed of triglycerides or cholesterol enveloped within protein
  - After formation they are absorbed into lacteals (blind lymphatic capillaries)
  - Transported in the lymph channels back into the CVS
  - Chylomicrons deliver lipids to diverse areas of the body
    - Liver—major delivery site; body's chemistry plant
    - Adipose tissue—storage of energy
    - Skeletal muscle—energy production
    - Cardiac and visceral muscle for energy production
- Three **broad categories of lipoproteins** produced in the liver for circulation in the CVS
  - **VLDLs** (very low-density lipoproteins)
    - Lipid delivery vehicles
    - Circulate in blood stream
    - Release triglycerides to cells—mostly adipose tissue
  - **LDLPs** (Low-density lipoproteins)—deliver cholesterol to cells
    - All cells integrate cholesterol into cell membrane—providing stability
    - Delivery of cholesterol to specialized cells that produce steroid hormones
      - Adrenal cortical hormones—aldosterone, cortisol, androgenic hormone
      - Ovaries—estrogen and progesterone
      - Testes—testosterone

BRIDGE to chapter 14, CVS, lymphatics, page 199–200

- **HDLs** (high-density lipoproteins)—transport lipids from cells to liver
  - HDLs are made in the liver
  - Circulate throughout the body
  - Conceptually they are empty lipoproteins
  - Since they are "empty" they can gobble up cholesterol
    - Cholesterol from peripheral tissues
    - Cholesterol from cholesterol plaques in arterial walls
  - Make cholesterol available to cells synthesizing steroid hormones
  - Carry cholesterol back to liver that converts excess cholesterol into bile salts for excretion from the body.
  - Help to regulate cholesterol levels. Desirable to have high HDL levels
  - **Remember** that most cholesterol in the blood stream is made in the liver. It does not come from the food we ingest.

Review Questions

- What are the three primary nutrients?
- What is the difference between essential and nonessential nutrients?
- Where is calcium essential for normal physiological function?
- What are chylomicrons?
- What are the differences between VLDLs, LDLs, and HDLs?
- Why is HDL beneficial?
- Why is cholesterol necessary for life?
- How does the body produce energy?
- How does the body transport lipids?

Activities

- With another A&P student, or someone who's completed A&P, discuss in detail the homeostatic balance of blood glucose and blood calcium.
- Write out an explanation of why HDL is beneficial to human health.

Vocalization Is Essential to Learning!

Churn to Learn!

Teach to Learn!

CHAPTER TWENTY

Fluids and Electrolytes

Acid/Base Balance

TOES (TOPICS OF EMPHASIS)

- Fluid compartments in the body
- Fluid exchange
- Fluid intake and output
- Acid/base balance
- Chemical and physiological buffering systems
- Disorders of acid/base balance

Roughly two-thirds of the human body's mass (weight) is composed of water. Normal human physiology is dependent on fluid homeostasis. The body has a variety of mechanisms allowing it to maintain fluid and acid/base balance.

Fluid Compartments

By now you have your head around the concept of one thing having two divisions (rule of two!). So, you can guess the body has two fluid compartments. The body's fluids are contained in one of two areas.

- **Intracellular fluid** (**ICF**)—Remember that cells are the basic structural and functional unit of the body, so you'd expect this would be one of the two fluid compartments containing fluid inside cells
    - Approximately 70 percent of body fluids are inside cells
    - SPPLBL is the barrier controlling fluid movement into and out of cells
- **Extracellular fluid** (**ECF**)—If it's not inside the cells, it'll be outside the cells.
    - **Interstitial** (**ISF**)—Fluid between the cells (two-thirds of extracellular fluid)
    - **Intravascular** (**IVF**)—Fluid inside the CVS vessels = **plasma** (one-third of extracellular fluid)

✔ **CONCEPT!** **Rule of thirds for body fluids**: ICF = 2/3 and ECF = 1/3 of body fluids. Then ECF breaks down as ISF = 2/3 and IVF = 1/3. This makes it easier to remember the distribution of body fluids.

Figure 20.1 shows the rule of thirds in graphic form to make it easier to retain. Pictures are worth a thousand words.

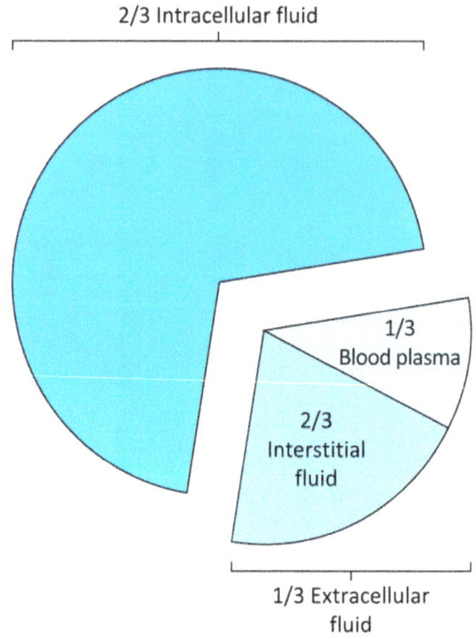

FIG. 20.1 Rule of thirds for body fluids.

There is constant movement of fluid (H_2O) between these compartments based on the relative osmolarity (concentration of solute) of each fluid compartment. When fluid in one compartment is hypotonic or hypertonic compared with another compartment, water will immediately move by osmosis toward the hypertonic solution. **Remember, solutes suck!** Salt, or sodium chloride is the primary solute involved in fluid movement in the body.

BRIDGE to chapter 7, page 72 for a review of osmosis!

Water will diffuse toward the compartment with the higher solute concentration (hypertonic) until water concentration is equal between the compartments. This is possible because the cell membrane is an SPPLBL. It's semipermeable! Water freely moves from one compartment to the other because of the way the SPPLBL is structured. Remember, things are structured to way they are because of what they do (**FFF**).

Also remember it's the **lymphatic system** that is responsible for the collection of excess interstitial fluid, draining the ISF into the left and right subclavian veins through the thoracic duct and the right lymphatic duct.

BRIDGE to chapter 14, CVS, lymphatic system, page 199–200.

Figure 20.2 graphically demonstrates water movement from hypotonic solution toward hypertonic solution. Notice the water is moving back and forth from ICF through ISF and into and out of IVF depending on the concentration of solute and the different osmotic gradients.

When a person is **dehydrated**, their plasma is hypertonic and water

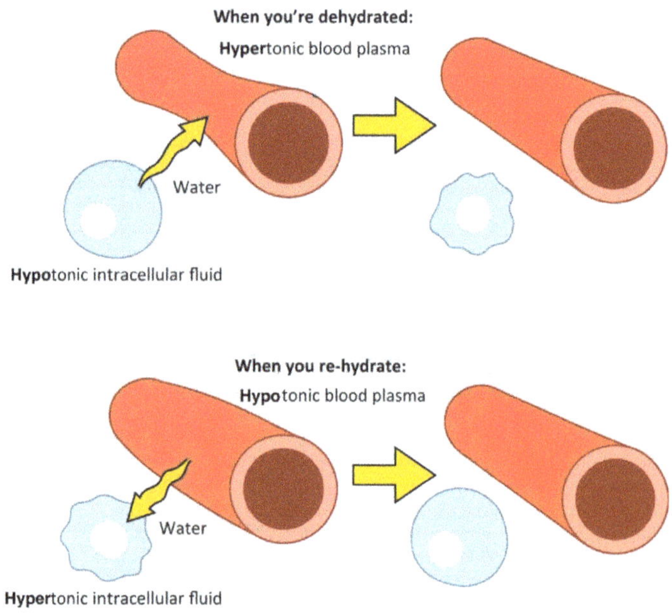

FIG. 20.2 Movement of water in dehydration and rehydration.

has a net movement from cells through the interstitial space into the intravascular space, attempting to reach a point of equilibrium. As a result, the cell shrinks.

When a person is **rehydrating,** the plasma becomes hypotonic and water will have a net movement from the intravascular space through the interstitial space into the cell. The cell will swell. Over-hydration (water intoxication) would cause extreme swelling of cells and potentially hazardous consequences, even death.

Figure 20.3 shows typical intake and outflow of fluid (water). Obviously, what we drink and the food we eat provide most of our water intake. This is called preformed water. A small amount of water is metabolic. It is produced as one of the two waste products of CR and from dehydration synthesis reactions.

Fluid output includes the following:

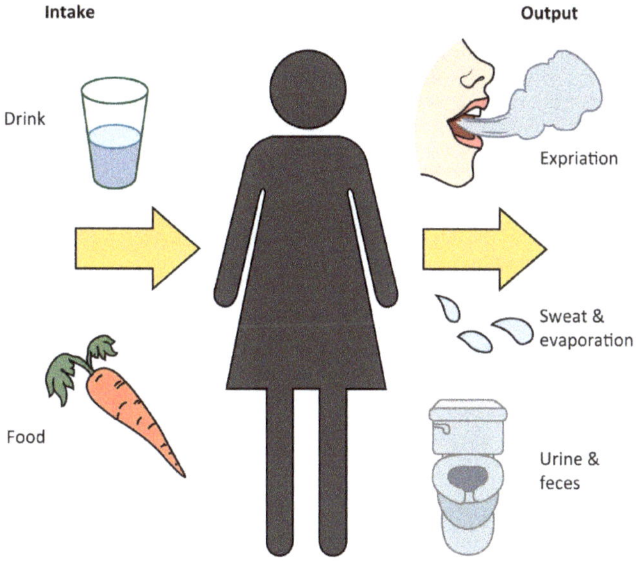

FIG. 20.3 Fluid intake and output.

- **Obligatory**—loss of water that always occurs
  - Loss occurs regardless of state of hydration
  - Loss though breathing and skin—insensible loss
- **Facultative**—controlled water loss
  - Regulation of amount of water loss
    - Hormonally regulated (ADH)
    - Depends on the amount of hydration
- **Sensible** water loss is measurable
  - Urine
  - Feces
- **Insensible** water loss is not measurable
  - Loss through breathing—expiration
  - Skin loss—sweating

BRIDGE
to Chapter 5, Chemistry, page 48, to review dehydration synthesis and hydrolysis reactions

NOTES

Acid/Base Balance and Buffering Systems

BRIDGE
to Chapter 5, chemistry, page 49 to review the concept of pH.

The normal range of pH (acidity) in the body is 7.35 to 7.45. This is a narrow range, and it is essential for the body to maintain that range. The body has three buffer (balancing) systems in place to maintain the homeostatic balance of hydrogen ion (H^+) concentration in the fluid compartments, particularly the IVF (intra-vascular fluid).

The body responds by compensating with specific buffering systems, either chemical or physiological, anytime the plasma pH goes outside (above or below) the normal pH range.

Buffering systems help to cushion or balance the level of acidity in cells and plasma. **Chemical buffering systems** are located within the cells or in the plasma and buffer within seconds or minutes. **Physiological buffering systems** include the respiratory system that buffers pH within minutes and the kidneys (urinary system) that buffer within hours or days. These systems are working constantly and collaboratively to maintain pH homeostasis.

✔ **CONCEPT!** There are two buffering systems—**chemical**, within the cells and blood, and **physiological**, in the lungs and kidneys.

There are three specific chemical buffering systems of importance. These chemical buffering systems minimize pH changes by temporarily binding and releasing hydrogen ion within seconds.

- **Phosphate buffering**—buffers within the cells
  - Involves hydrogen phosphate molecules
    - Inorganic molecule
    - Phosphorus plus four oxygen atoms and hydrogen
    - Gives up or receives H^+ to buffer acidity
- **Protein buffering**—occurs both within cells and in blood
  - Involves protein molecules—contain NH_2 and COOH groups
  - NH_2 buffers acids and COOH buffers alkaline solutions
- **Bicarbonate buffering**—involves carbonic acid and bicarbonate

$$CO_2 + H_2O \leftrightarrow H_2CO_3 \leftrightarrow HCO_3^- + H^+$$

  - Same formula as discussed in chapter 16
  - Reaction goes both ways
  - Driven by reagent concentration
  - Increase H^+ (acidic) pushes reaction to the left
  - Decrease H^+ (alkaline) pulls reaction to the right

The **physiology buffering systems** maintain pH by eliminating excess acid or base from the body through the respiratory and urinary systems.

- **Respiratory system buffering**—involves the lungs and ventilation; works within minutes

$$CO_2 + H_2O \leftrightarrow H_2CO_3 \leftrightarrow HCO_3^- + H^+$$

  - Involves this same carbonic acid formula
  - If body is acidotic (low pH) it will compensate by increased respiratory rate—hyperventilation which:
    - Decreases CO_2 levels
    - Draws reaction to the left
    - Results in fewer hydrogen ions
    - Lower acidity—returns to homeostatic pH balance
  - If body is alkalotic (high pH) body will compensate by decreased respiratory rate—hypoventilation
    - Increases CO_2 levels
    - Pushes reaction to the right
    - Results in more hydrogen ions
    - Higher acidity—returns body to homeostatic pH balance
- **Urinary system buffering**—kidneys; works within hours or days
  - If body is acidotic (low pH) body will compensate by the following kidney functions:
    - Increased excretion (peeing away) of H^+
    - Reabsorption of bicarb (HCO_3^-)
    - Higher pH = lower acidity, returning body to homeostatic pH balance
  - If body is alkalotic (high pH) body will compensate by the following kidney functions:
    - Decreased excretion of H^+
    - Secretion of bicarb (HCO_3^-)
    - Lower pH = higher acidity, returning body to homeostatic pH balance

Once again, it's helpful to understand the body is highly dynamic and the compensation processes are constantly in flux, changing and working together to maintain acid/base homeostasis.

NOTES

Acid/Base Imbalances

There are two types of acid/base imbalance, acidosis (lower pH), and alkalosis (higher pH). Once again, we see one thing divided into two subcategories! Rule of two!

✔ **CONCEPT** Rule of two—two types of acid/base imbalances, acidosis and alkalosis which each have two types, metabolic and respiratory.

- **Metabolic imbalances**—hydrogen ion fluctuation caused by body processes not involving the lungs
  - **Metabolic acidosis**—increased hydrogen ions or decreased bicarb
    - Body produces too many hydrogen ions
      - Ketoacidosis in diabetes
      - Increased lactic acid from CR
      - Excessive intake of ETOH (booze) leading to increase in acetic acid
    - Decreased renal function—allowing H^+ to accumulate
    - Severe diarrhea—increased loss of bicarb
  - **Metabolic alkalosis**—decreased hydrogen ions or increased bicarb
    - Loss of hydrogen ions by excessive vomiting
    - Loss of hydrogen ions by excessive nasogastric suctioning in hospital
    - Loss of acids by the kidney caused by overuse of diuretics
    - Increased alkaline input from excessive use of antacids
  - **Respiratory acidosis**—increased CO_2 levels
    - **Hypoventilation** and accumulation of carbon dioxide because of one or more of the following:
      - Trauma to the respiratory centers of the brain
      - Disorders of the nerves and muscles of breathing
      - Something blocking the airways
      - COPD—chronic obstructive pulmonary disease
        - Decreased gas exchange
        - Emphysema
        - Pulmonary edema
        - Chronic bronchitis
        - Fibrotic thickening of alveolar membrane
  - **Respiratory alkalosis**—decreased levels of CO_2
    - **Hyperventilation** and lower levels of carbon dioxide
      - Severe anxiety

- Not getting enough oxygen to tissues
    - High altitude
    - Severe anemia
    - Congestive heart failure
    - Low blood pressure (BP)
- **Aspirin overdose**—stimulates respiratory centers in the brain, causing hyperventilation

In any condition of acidosis or alkalosis, the body will attempt to compensate, utilizing both the chemical buffer systems (phosphate, protein, and bicarbonate) and the physiological buffer systems (respiratory and urinary).

Acid/base balance is a challenging concept to grasp, so novice students should not be discouraged if they struggle to understand it. Insight will take time, lots of repetition, and the application of **REVV CCCAMM!** So, be patient as you strive to understand this important concept.

NOTES

Review Questions

- What are the three fluid compartments of the body?
- What is the difference between obligatory and facultative fluid output?
- What's the difference between sensible and insensible fluid loss?
- What are the three types of chemical buffering systems?
- What is the difference between metabolic and respiratory pH imbalance?
- Why is the bicarbonate formula essential to understanding acid/base balance?
- What are three causes of metabolic acidosis?
- What are three causes of respiratory alkalosis?

Activities and Coloring

- Diagram and color what happens to a human RBC when it is placed in hypotonic (distilled water), isotonic (0.9 percent salt), and hypertonic (5 percent salt) solutions.
- Write out the bicarbonate buffer formula and explain to another person how changing substrate concentrations causes the reaction to flow either direction (left or right).

- Write out a comparison of respiratory and metabolic acidosis and alkalosis.

FIGURE CREDITS

Figure 20.1: Adapted from McGraw-Hill Education.
Figure 20.2: Adapted from McGraw-Hill Education.
Figure 20.3: Adapted from McGraw-Hill Education.

Vocalization Is Essential!

Churn to Learn!

Teach to Learn!

CHAPTER TWENTY ONE

Reproductive System

Continuation of the Human Race

TOES (TOPICS OF EMPHASIS)

- Midsagittal section of the female pelvis
- Midsagittal section of the male pelvis
- Conceptual function of the reproductive systems
- Similarities and differences of female and male systems
- Exocrine and endocrine functions
- Male testosterone levels
- Female endocrine hormone fluctuations

Reproductive System

The reproductive system allows for the continuation of the human species. As vital as this system is, at an introductive level it is many times de-emphasized. After all, most introductory courses are packed with details about the other eleven systems. But some basic knowledge of the structure (and function) of the reproductive system is important.

Anatomically, the novice student needs to know the structures seen in a midsagittal section of both the female and the male pelvis. (See Figures 21.1 and 21.2.)

These are the anatomical structures of importance in the female.

- Pubic bone—symphysis pubis
- Urinary bladder
- Urethra
- Clitoris
- Labia minora
- Labia majora
- Ovary
- Sigmoid colon
- Uterus

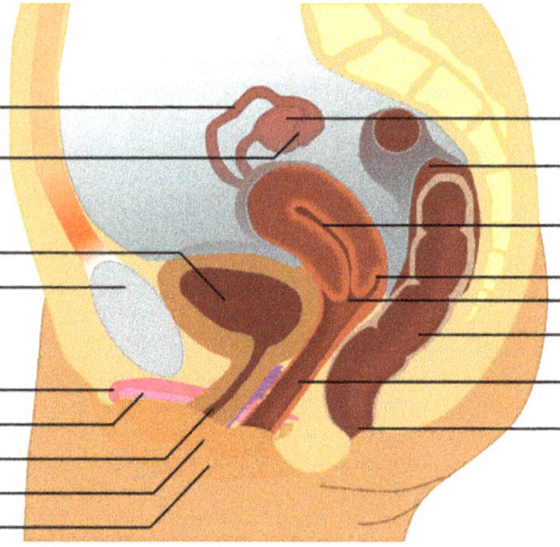

FIG. 21.1 Midsagittal Section of Female Pelvis.

- Cervix
- Vagina
- Anus

NOTES

These are the anatomical structures of importance in the male.

- Pubic bone—symphysis pubis
- Urinary bladder
- Prostate gland
- Urethra
- Penis
- Sigmoid colon
- Rectum
- Anus
- Testicle
- Epididymis
- Vas deferens—spermatic duct

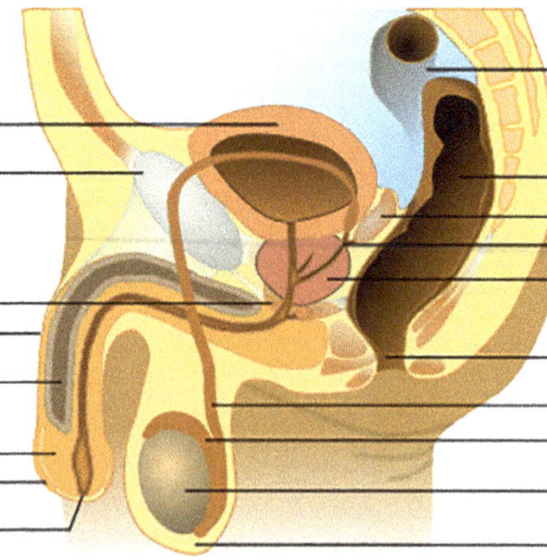

FIG. 21.2 Midsagittal Section of Male Pelvis.

NOTES

The male and female reproductive systems are designed to allow for the fertilization of an oocyte by sperm. This results in the formation of a single cell zygote that grows into a complete human organism through the process of mitosis. The zygote implants in the highly vascular endometrium of the uterus where it develops into a human organism.

The reproductive systems in both the male and female are essential for the continuation of the human species. But it is the only system not necessary to sustain the life of an individual organism. Life wouldn't be as interesting, or as much fun, or as confusing at times, but as important as it is, the reproductive system is the only non-necessary body system. Although in the male, some structures of the reproductive system are shared by the urinary system.

SS A&P does not go into any significant detail regarding reproductive function. However, it is essential for the A&P student to recognize the complexity of these systems, especially the female reproductive system. Although obviously different, both female and male systems share similarities.

Similarities of Female and Male Reproductive Systems

- Both have primary reproductive organs called **gonads** which produce sex cells
  - Ovaries in the female—eggs or oocytes
  - Testes in the male—sperm
- Both produce endocrine hormones (so they are part of endocrine system) but also have an exocrine function
  - Combined endocrine and exocrine function in female
    - Exocrine—egg production and movement through Fallopian tubes
    - Endocrine—estrogen and progesterone (steroid-based hormones)
  - Combined endocrine and exocrine function in male
    - Exocrine—sperm production and movement through vas deferens
    - Endocrine—testosterone (steroid-based hormone)
  - Both sexes also exhibit accessory reproductive organs
    - Female vagina—ducts to carry egg from ovary, through fallopian tubes to uterus
    - Male penis —ducts carry sperm from testes to the outside

NOTES

Male Reproductive Function

The production of **LH** (luteinizing hormone) and **FSH** (follicle stimulating hormone) by the anterior pituitary gland after the first decade of life, brings the male into puberty. Once he reaches puberty, testosterone levels dramatically increase. This creates secondary sexual changes such as lower voice, beard growth, and increased muscle mass.

The stimulation of pituitary hormones also initiates the production of sperm by the **Sertoli cells** of the testes. Sperm is stored and matures in the epididymis until a time of ejaculation when the sperm is forced to the outside by the contractions of visceral muscle in the male pelvis and penis. Erection and ejaculation in the male are controlled by the ANS (Autonomic Nervous System). This is a case where the sympathetic and parasympathetic responses are synergistic and not antagonist as they usually are. Sympathetic stimulation facilitates erection whereas parasympathetic stimulation causes ejaculation. The average ejaculation volume is 5 cc and contains approximately one-half million sperm. This is an example of redundancy that helps to guarantee fertilization will occur during intercourse.

Sperm and testosterone production are under hormonal endocrine regulation involving classical negative feedback loops. The novice student needs to focus on the general concept of male reproductive endocrine control rather than struggle to remember all the details.

Testosterone levels in the male remain constant for decades and then start to decline in the sixth and seventh decades of life. But there is no periodic fluctuation of hormones as there is in the female. This makes the male relatively simplistic and straightforward physiologically.

Female Reproductive Function

Production of **LH** and **FSH** brings the female into puberty. Once she reaches puberty the ovary starts to develop and ovulate eggs. The female is born with all the eggs she'll ever have, approximately one million for each ovary for a total of two million eggs. Only a few hundred of those eggs ever reach maturity and ovulate.

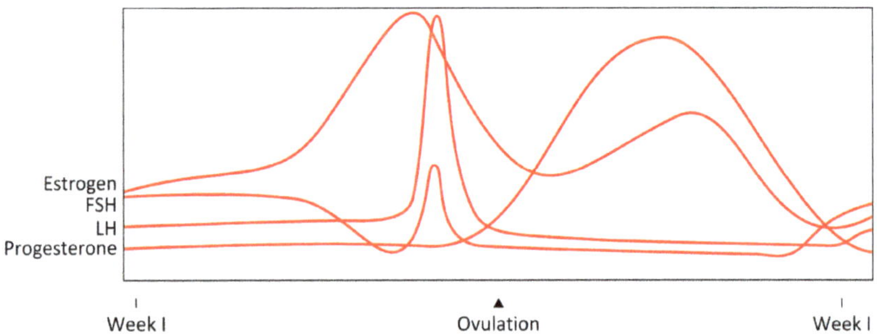

FIG. 21.3 Female Hormonal Fluctuations During Menstrual Cycle.

Figure 21.3 shows the complexity of female reproductive

physiology. At an introductory level it is not necessary to memorize or understand the details of female physiology. What is essential is a **conceptual understanding** of the significant complexity of female reproductive physiology.

Female reproductive regulation involves classic negative feedback loops just as it does in the male. Focus on the concept of reproductive control more than any of the smaller details contained in the traditional A&P text or some of the online sources you're utilizing as reference resources.

A picture is worth a thousand word and Figure 21.3 says it all about the complexity of female reproductive physiology. Notice the fluctuations during the menstrual cycle with peaks of LH, FSH and estrogen at the time of ovulation. Progesterone then peaks late in the cycle. When the progesterone level falls at the end of the cycle, menstruation occurs.

Review Questions

- What are the similarities of female and male reproductive systems?
- What anterior pituitary hormones are essential in the regulation of both female and male systems?
- List the steroid-based hormones involved in both the male and female.
- What lipid molecule is necessary as the raw material for both male and female sex hormones?
- The myometrium of the uterus is composed of what specific tissue?
- What is the sex organ of the male?
- What is the sex organ of the female?
- What are the differences between the female and male systems?

Activities and Coloring

- Explain the similarities and differences between the female and male reproductive systems to a classmate or family member.
- Draw and color a line diagram of the four female hormones (two pituitary and two ovarian) showing fluctuations during the menstrual cycle.
- Search online for an unlabeled midsagittal section of both the male and female pelvis, print them out, color and label them. Say the names of each structure out loud as you fill in the label and/or as you color.

FIGURE CREDITS

Visualize and Vocalize!

Draw It to Know It!

Churn to Learn!

Teach to Learn!

EPILOGUE

It has been a long and sometimes challenging journey through all the chapters of *SS A&P*, so an epilogue seems appropriate.

You are now able to integrate anatomy and physiology to gain a complete and holistic perspective of the human being. You can understand how anatomy and physiology are inseparable, constantly working together to maintain homeostasis.

As you move forward with your academic endeavors and your healthcare career, *SS A&P* will provide an effective review tool. It will be an old friend preparing you for pathophysiology or clinical rotations or certification exams. As a licensed health care provider, you will find it a useful review and refresh tool!

The concepts in *SS A&P* will renew your insight into this marvelous human body we all occupy. This will allow you to bring greater value to those you are helping in the wide spectrum of clinical settings making up the health care world.

Remember, this will be a long journey, requiring you to be a **lifelong learner**!

Keep churning and learning!

:o)

GLOSSARY

A

Abdominal regions—the nine surface areas of the abdomen
Abdominal quadrants—the four surface quadrants of the abdomen
Accommodation—process of focusing light on the retina when changing from close to distant vision or vice versa
Acetylcholine—ACh—common neurotransmitter; found at the NMJ of skeletal muscle
Acetyl CoA—molecule that bridges from glycolysis into the Kreb's cycle in CR (stargate molecule)
Action potential—depolarization/repolarization of an axon; causes an electrical impulse
Active transport—moves solute molecules against a concentration gradient; requires energy (ATP)
Adaptive immunity—provides antibody response to specific antigenic stimulation
Aerobic—processes that require oxygen
Agonist—muscle that is a prime mover, creating a specific action
Anabolism—chemical reactions that build larger molecules from smaller particles
Anaerobic—process that does not require oxygen
Anatomical plane—imaginary surface that slices the body into sections
Anatomical position—the standard reference body position used in A&P
Anatomical regions—specific names for the various areas of the body (e.g., cervical or axillary)
Anatomical terminology and nomenclature—the language of anatomy and physiology
Anion—a negatively charged particle
Antagonist—muscle whose movement opposes the action of an agonist
Atlas—first cervical vertebrae; C-1
Atom—the smallest particle that exhibits the chemical properties of an element
Axis—second cervical vertebrae; C-2

B

Basal cell carcinoma—cancer originating from the basilar layer; least dangerous type of skin cancer
Biochemical—large, organic macromolecules like proteins, carbohydrates, lipids and nucleic acids
Body cavities—cavities in the body that are either open or closed to the outside
Body systems—twelve systems in the body that produce specific functions, including gas exchange, digestion, and blood circulation
Boxer's muscle—serratus anterior; overdevelops in boxers
Boyle's law—pressure and volume are inversely related in gas-filled chambers
Buffering systems—processes that help to balance acidity in the body
Bulk transport—vesicular transport

C

Carbohydrates—saccharide (mono, di, and poly), which makes up foods like fruit juices pasta and potatoes

Cardiac cycle—six superimposed graphs showing all the events occurring in the heart in one heartbeat
Cardiac muscle—muscle found in the walls of the heart; built in rhythmicity
Cardiovascular system (CVS)—body's transportation system
Catabolism—chemical reactions that break large molecules into smaller particles
Cation—a positively charged particle
Cell biology—the study of the cell structure, intracellular organelles, and cellular functions
Cellular immunity—adaptive immunity requiring direct contact between pathogen and T-cells
Cellular metabolism (**CM**)—the sum of all metabolic (catabolism and anabolism) reactions in cells
Cellular respiration (**CR**)—the process of energy production (catabolic)
Chemistry—the science dealing with the composition and properties of substances and how they interact with each other
Chylomicrons—lipoprotein necessary for the absorption of fat from the small intestine
Citric acid cycle (Kreb's cycle)—stage three in the CR process
Colloid—cloudy mixtures of large particles that remain mixed when not in motion
Conducting zone—division of respiratory system that conducts air in and out of lungs, ventilation
Connective tissue—a primary tissue that connects and supports
Cori cycle—process of converting lactate into glucose in the liver during anaerobic activities
Covalent bond—the strongest chemical bond where atoms share one or more electrons
Cranial nerves—peripheral nerves (PNS) that arise from the area of the cranium

D

Dehydration synthesis—creation of larger molecules involving the joining of two smaller molecules and the production water
Dermis—true skin; two layers: reticular and papillary
Diaphysis—shaft of long bone
Diffusion—random movement of molecules from higher to lower concentration; simple or facilitated
DNA—Deoxynucleic acid double-helix molecule in the cell nucleus containing genetic codes

E

EKG (ECG)—electrocardiogram; graph showing the electrical activity of the heart
Electron transport chain (ETC)—fourth stage in CR; produces majority of ATP
Enzyme—molecules (proteins) that catalyze (speed up) chemical reactions
Epithelium—one of the four primary tissues that line and cover structures in the body
Endergonic reaction—chemical reaction that uses energy
Endocrine system—secondary communication system utilizing chemical hormones
Endocytosis—process of cells ingesting substances from outside the cell to inside the cell
Endomysium—loose areolar connective tissue wrapping around muscle cells
Endoneurium—loose areolar connective tissue wrapping around individual neuronal axons
Endoplasmic reticulum (**ER**)—manufacturing site of proteins (rough) or lipids (smooth)
Epidermis—the layer of skin above the dermis; made of epithelial tissue
Epimysium—dense regular connective tissue sheath surrounding muscle fascicles
Epineurium—connective tissue sheath surrounding nerve fascicles
Epiphyseal line—fused growth plate of long bone in adult
Epiphysis—end of a long bone
Epiphysis plate—growth area of long bone in youngsters
Epithelium—primary tissue that covers and lines
Exergonic reaction—chemical reaction that gives off energy
Exocytosis—process of cells releasing substances from intracellular space to extracellular space

F

FEV1—the amount of air that can be forced out of the lungs in one second
FFF (form follows function)—the basic underlying concept of *Super Simple A&P*. Structures in the body are shaped the way they are because of what they do
Fascicles—groupings of muscle cells and or neuron axons; surrounded by perimysium or perineurium
Fluid compartments—areas of the body containing fluid (water); intracellular and extracellular
Follicle-stimulating hormone (FSH)—endocrine hormone produced by the anterior pituitary gland; essential in sexual development and function in both make and female
Foramen—opening in a structure, like bone
Fright muscle—platysma muscle; only muscle with minimal to no boney attachments running skin to skin

G

Glomerular filtration rate (GFR)—standard measurement of the amount of filtrate produced in one minute by the kidneys
Glucose—monosaccharide molecule that is the primary nutrient used in the production of energy
Glycolysis—a chemical process involving the breakdown of glucose into pyruvate
Gyrus—ridge in the cerebrum

H

H_2O sheet—an innovative flowchart displaying anatomy, physiology, and pathophysiology on a single sheet; shows organizational hierarchy, biochemicals, primary tissues, body systems, and disease processes
Histology—the study of tissues
Homeostasis—Dynamic constancy (DC); the body's ability to maintain a relatively constant internal environment (within an upper and lower limit) when the external environment is changing
Hormonal endocrine control—trophic hormones target other endocrine glands
Humoral endocrine control—glands secrete in response to change in the level of molecules in the blood, e.g. glucose and calcium
Humoral immunity—type of defensive response producing antibodies; B cells
Hydrogen bond—chemical bond caused by weak attraction between polar molecules
Hydrolysis—involves the breakdown of larger molecules into smaller particles with the breakdown of water
Hyoid bone—unique, freestanding bone without any boney articulation

I

Innate immunity—providing general immunity (barrier and non-barrier) to a variety of hazards
Integumentary—skin; largest organ of the body that is also a system; contains and protects
Ionic bond—electrons are completely given up by one atom and absorbed by another atom
Ions—an electrically charged particle
Isometric contraction—contraction of muscle without movement
Isotonic contraction—contraction of muscle that creates movement
Isotope—an atom with the addition of one or more neutrons in the nucleus

J

Joint actions—movement around joints (e.g., flexion, extension, abduction, adduction, etc.)

K

Kreb's cycle (citric acid cycle)—stage three in the CR process
Kyphosis—exaggerated thoracic curve of the spine; abnormal and pathological

L

Limbic system—multiple structures in the cerebrum and diencephalic areas that process emotions and memory
Lipids—water-soluble fat molecules like triglycerides
Lordosis—exaggerated lumbar curve of spine; abnormal but may be non-pathological (3rd trimester of pregnancy)
Luteinizing hormone (LH)—endocrine hormone produced by the anterior pituitary gland; essential in sexual development and function.
Lymphatic system—part of the immune system with a secondary function of resorption of interstitial fluid

M

Macromolecules—large biochemicals that make up the human body; organic (containing carbon) structures including carbohydrates, proteins, lipids, and nucleic acids (DNA, RNA)
Major chemical elements—those elements that make up more than 98 percent of the body's weight
Malignant melanoma—arises from melanocytes in the basilar layer of the skin; deadly type of cancer
Medical terminology—the study of how medical (including A&P) words are structured with word roots, suffixes, and prefixes
Meiosis—produces sex cells (eggs, sperm) with haploid number (23) of chromosomes
Meninges—connective tissue layers covering the brain and spinal cord (CNS)
Mesentery—specific folding of the serous membrane lining the abdominopelvic cavity which supports intra-abdominal organs
Messenger RNA (mRNA)—carries genetic code from nucleus to cytoplasm in protein synthesis (PS)
Metabolic imbalances—types of acid/base fluctuation; acidosis/alkalosis, metabolic/respiratory
Metaphysis—area between the diaphysis and epiphysis of a long bone
Minor chemical elements—those elements that make up less than 1 percent of body weight
Mitochondria—intracellular organelle that produces energy (ATP); the cellular energy factory
Mitosis—cell division where mother cell divides into two identical daughter cells (diploid—46 chromosomes)
Motor unit—one motor somatic neuron and all the muscle fibers it innervates
Muscle fiber—a muscle cell; bundles of myofibrils; surrounded by endomysium
Muscle twitch—a single brief contraction and relaxation of a muscle
Muscular tissue—one of the four primary tissues that produces movement in the body
Myofibrils—bundle of myofilaments
Myofilaments—proteins; actin and myosin

N

Negative feedback loop—feedback system that reverses the original stimulus
Nephron—functional and structural unit of the urinary system
Neuroglial cells—six different types of cells supporting neurons in the nervous system (CNS and PNS)
Neuromuscular junction (NMJ)—area where motor somatic neuron connects to skeletal muscle
Neuron—structural and functional unit of the nervous system; transmits electrical impulse
Nervous endocrine control—glands release hormones in response to nervous stimulus
Nervous tissue—a primary tissue that provides communication (electrical) inside the body
Non-polar covalent bond—chemical bond where electrons are shared equally, thus no electrical polarity

O

Organic chemicals—any molecule that contains the element carbon
Organizational hierarchy—the body is organized by starting with tiny particles like subatomic particles (protons, neutrons, and electrons) and combining them repetitively to form cells, tissues, organs, organ systems, and finally the human organism; conceptually, humans are organized from small to large
Osmosis—diffusion of water across an SPPLBL toward a concentration of solute
Osteoblast—bone-building cell
Osteoclasts—bone-resorbing cells (chew bone)
Oxidation/reduction reaction (redox)—a chemical reaction where one molecule loses electrons and the other molecule gains electrons

P

Pathophysiology—the study of how disease affects normal physiological function
Perimysium—connective tissue wrapping around muscle fascicles
Perineurium—connective tissue wrapping around axon bundles, forming a fascicle
Periodic table—tabular chart of all chemical elements arranged by atomic number, electron configuration, and recurring chemical properties
Periosteum—tough sheath of dense irregular connective tissue covering the outside of bone
pH scale—chemical scale showing the range of acidity in solutions
Polar covalent bond—chemical bond where electrons are shared unequally, causing electrical polarity
Positive feedback loop—feedback system that amplifies the original stimulus
POSS (physiology on a single sheet)—A flowchart detailing the ongoing process of ingesting and digesting food (carbohydrates, proteins, and fats), circulating the products of digestion (monosaccharides, amino acids, and fatty acids) in the blood stream to cells. Cells then utilize those smaller molecules to either make larger molecules, like proteins (e.g., insulin or thyroid hormone) or to create energy through cellular respiration
Primary tissues—The human is made up of four primary tissues, epithelium, connective, muscular, and nervous. Organs are a combination of those four primary tissues
Protein synthesis—the process of manufacturing protein molecules
Proteins—polypeptide chains made from amino acids

R

RNA—Ribonucleic acid - single helix molecule produced by the nucleolus (ribosomes)
Redox reaction (oxidation/reduction reaction)—a chemical reaction where one molecule loses electrons and the other molecule gains electrons (OIL RIG)
Reflex arc—rapid, involuntary response system operating at a spinal cord level
Respiratory zone—division of the respiratory system where gas exchange takes place
Ribosome—intracellular organelle that produces proteins (protein factory)
Ribosomal RNA (rRNA)—nucleic acid making up ribosomes in the cytoplasm and ER

S

Scientific method—a standardized process that is the basis of the development and evolution of modern-day sciences. It involves six steps: observation, question, hypothesis, experimentation, data analysis, and conclusions
Scoliosis—abnormal lateral curvature of the spine; pathological
Sebaceous glands—skin glands that produce oily, waxy substances
Serous membranes—specific membrane that lines closed body cavities
Skeletal muscle—voluntary or striated muscle that produces movement of the skeleton
Sliding filament theory—current theory of how muscle contracts
Solute—solid particles that will dissolve in a solvent
Solution—mixture where solutes disperse (dissolve) to make a clear mixture
Solvent—substances that dissolve solutes (solid)
Sphincter—circles of muscle (usually visceral) that surround tubular structures
SPPLBL—Semipermeable, phospholipid, bi-layer; cell membrane
Squamous cell carcinoma—originates from keratinocytes in the stratum spinosum
Stargate molecule—Acetyl-CoA in CR; common entry point for all three nutrition sources
Steroid hormones—lipid-based (cholesterol) hormones produced in adrenal cortex, ovaries, and testes
Subatomic particles—particles that make up atoms: protons, neutrons, and electrons
Sulcus—depression (valley) in bone or the brain
Surface tension—attraction of polar water molecules through hydrogen bonding
Suspension—large solute particles are held suspended in the mixture by the motion of water
Sweat glands—two types: eccrine (merocrine) or apocrine
Synapse—the structure where two neurons join
Synergist—muscle that assists the action of an agonist
Synovial joint—most common joint in the body; freely moveable

T

Tailor's muscle—sartorius—longest muscle in the body—overdeveloped in old-time tailors
Ten keys to learning—REVV CCCAMM = repetition, emotion (+ or -), visualization, vocalization, concepts, communication, collaboration, association, movement, and memorization
Tidal volume—amount of air moved during ventilation at rest; approximately 500 cc
Tonicity—the ability of a solution to change the volume of tone of a cell by adding or subtracting water through osmosis
Transcription—copying of genetic code from DNA to mRNA in the nucleus
Transfer RNA (tRNA)—transports amino acids to ribosomes in protein synthesis (PS)
Translation—process of stringing amino acids together to make polypeptide chains (proteins)

V

Vesicular transport— cell membrane transport of large substances or large amounts of smaller molecules through membrane-bound sacs (vesicles); also called bulk transport
Visceral muscle—smooth or involuntary muscle found in the walls of viscera (guts)
Vital capacity—total amount of air that can be exchanged through one maximum inhalation and exhalation

W

Water—essential molecule of great importance in human physiology; solvent for human physiology
Water mixtures—formed by combining water (solvent) and other substances (solutes)

Z

Zygote—single cell fertilized egg; the result of fertilization

INDEX

A

Abdominal regions, 32
Abdominal quadrants, 31, 32, 40
Accommodation, 173
Acetylcholine, 141, 142, 143, 172
Acetyl CoA, 60, 61
Action potential, 142, 153, 169
Active transport, 74
Adaptive immunity, 209, 210, 213
Aerobic, 58, 60
Agonist, 135
Anabolism, 21, 55, 56
Anaerobic, 59, 145
Anatomical plane, 31, 33
Anatomical position, 38
Anatomical regions, 37
Anatomical terminology and nomenclature, 31
Anion, 43, 47
Antagonist, 135, 137, 183
Atlas, 108
Atom, 15, 27, 60
Axis, 108

B

Basal cell carcinoma, 92
Biochemical, 15, 49
Body cavities, 31, 34, 141
Body systems, 14, 23
Boxer's muscle, 139, 148
Boyle's law, 215, 218
Buffering systems, 248
Bulk transport, 75

C

Carbohydrates, 15, 227, 229, 241
Cardiac cycle, 191, 202
Cardiac muscle, 26, 83, 147
Cardiovascular system (CVS), 18
Catabolism, 55, 56
Cation, 43
Cell biology, xxvii, 65
Cellular immunity, 213
Cellular metabolism (CM), 57
Cellular respiration (CR), 21, 57
Chemistry, 41, 52, 248
Chylomicrons, 243
Citric acid cycle (Kreb's cycle), 58, 61
Colloid, 47
Conducting zone, 215, 217
Connective tissue, 16, 81
Cori cycle, 59, 60
Covalent bond, 44
Cranial nerves, 153, 161

D

Dehydration synthesis, 48
Dermis, 88
Diaphysis, 102
Diffusion, 70, 71, 72
DNA, 56, 63

E

EKG (ECG), 193, 202, 203
Electron transport chain (ETC), 58, 61
Endergonic reaction, 59
Endocrine system, 153, 181
Endocytosis, 75
Endomysium, 132, 141
Endoneurium, 156
Endoplasmic reticulum (ER), 66
Enzyme, 41, 51, 52
Epidermis, 88, 89
Epimysium, 132
Epineurium, 156
Epiphyseal line, 102
Epiphysis, 102, 105
Epiphysis plate, 105
Epithelium, 77, 216, 223
Exergonic reaction, 59

F

FEV1, 219
Fascicles, 132, 140
FFF (form follows function), 223
Fluid compartments, 245
Follicle-stimulating hormone (FSH), 256
Foramen, 102
Fright muscle, 139

G

Glomerular filtration rate (GFR), 237
Glucose, 238
Glycolysis, 58, 59
Gyrus, 159

H

H2O sheet, 168, 183, 225
Histology, xxvii, 3
Homeostasis, 99, 126
Hormonal endocrine control, 256
Humoral endocrine control, 182

Humoral immunity, 213
Hydrogen bond, 46
Hydrolysis, 48
Hyoid bone, 103

I

Innate immunity, 209, 210
Integumentary, 17
Ionic bond, 45
Ions, 69
Isometric contraction, 146
Isotonic contraction, 146
Isotope, 44

J

Joint actions, 78, 99

K

Kreb's cycle, 61
Kyphosis, 111

L

Limbic system, 160, 161
Lipids, 227
Lordosis, 111
Luteinizing hormone (LH), 256
Lymphatic system, 18

M

Macromolecules, 14, 15
Major chemical elements, 15
Malignant melanoma, 93
Medical terminology, 5
Meiosis, 69
Meninges, 158
Mesentery, 35
Messenger RNA (mRNA), 63
Metabolic imbalances, 250
Metaphysis, 102
Minor chemical elements, 42
Mitochondria, 56
Mitosis, 69
Motor unit, 141
Muscle fiber, 144
Muscle twitch, 145
Muscular tissue, 132
Myofibrils, 140
Myofilaments, 140, 148

N

Negative feedback loop, 99, 183, 185
Nephron, 235, 240
Nervous endocrine control, 157
Neuroglial cells, 156
Neuromuscular junction (NMJ), 171
Neuron, 154, 156
Non-polar covalent bond, 44

O

Organic chemicals, 49
Organizational hierarchy, 14, 41
Osmosis, 60, 72
Osteoblast, 99, 126, 128
Osteoclasts, 99, 126
Oxidation/reduction reaction (redox), 52

P

Pathophysiology, xxiii, 6, 19
Perimysium, 132
Perineurium, 156
Periodic table, 15, 42
Periosteum, 102
pH scale, 49
Polar covalent bond, 44
Positive feedback loop, 127, 184
POSS (physiology on a single sheet), 3
Primary tissues, 14
Proteins, 63
Protein synthesis, 26, 62

R

Reflex arc, 162, 165
Respiratory zone, 215, 217
Ribosome, 63, 64, 67
Ribosomal RNA (rRNA), 63
RNA, 63

S

Scientific method, 1, 2
Scoliosis, 111
Sebaceous glands, 88
Serous membranes, 35
Skeletal muscle, 59, 78, 99
Sliding filament theory, 144
Solute, 184, 236, 237
Solution, 246, 248
Solvent, 47
Sphincter, 224
SPPLBL, 242, 245
Squamous cell carcinoma, 92
Stargate molecule, 60, 61
Steroid hormones, 56, 185, 243
Subatomic particles, 15, 41
Sulcus, 105
Surface tension, 46
Suspension, 47
Sweat glands, 88, 91
Synapse, 171
Synergist, 135, 137
Synovial joint, 122

T

Tailor's muscle, 139
Ten keys to learning, 2, 10, 191
Tidal volume, 219
Tonicity, 73
Transcription, 55, 63
Transfer RNA (tRNA), 63
Translation, 55, 63

V

Vesicular transport, 75
Visceral muscle, 79, 147, 172
Vital capacity, 219

W

Water, 223, 236
Water mixtures, 47

Z

Zygote, 69, 239

www.ingramcontent.com/pod-product-compliance
Lightning Source LLC
LaVergne TN
LVHW070107021026
849160LV00026B/1697
* 9 7 8 1 7 9 3 5 1 7 7 3 9 *